ROYAL AIR FORCE

WHEELS OF THE RAF

VEHICLES OF THE FLYING SERVICES THROUGH TWO WORLD WARS

BRUCE ROBERTSON

PSL Patrick Stephens, Cambridge

Front endpaper *That the RAF has had more vehicles than aircraft is exemplified by this view of VC-10 XV109* Arthur Scarf VC *with over a dozen vehicles of various types attendant at Akrotiri in the 70s* (RAF NEAF).

Title page *Night refuelling scene of Britannia C1 XL658* Adhara (K.M. Robertson Collection).

Back endpaper *Karrier Bantam 2-ton airfield control truck (79 AE04) at runway edge controlling Vulcan B2 flying operations.*

First published in 1983

British Library Cataloguing in Publication Data

Robertson, Bruce
Wheels of the RAF.
1. Great Britain *Royal Air Force*—History
2. Trucks, Military—History 3. Vehicles, Military—History
I. Title
623.74'7 UG615

ISBN 0-85059-624-6

Photoset in 10 on 11 pt Times by Manuset Limited, Baldock, Herts. Printed in Great Britain on 115 gsm Fineblade coated cartridge by R.J. Acford Ltd, Chichester, and bound by The Garden City Press, Letchworth, Herts, for the publishers, Patrick Stephens Limited, Bar Hill, Cambridge, CB3 8EL, England.

Contents

RAF armour then and now. **Below** *a convoy of Morris four- and six-wheeled troop-carrying tenders escorted by RAF Rolls-Royce armoured cars in the 30s and* **above** *a review of RAF Regiment AFVs in the 80s with a Scorpion armoured reconnaissance vehicle leading three Spartan armoured personnel carriers* (MoD H1720 and RAF Regiment).

Introduction

Throughout the history of the Royal Air Force it will be found that its establishment of vehicles has exceeded that of its number of aircraft. Moreover, there has always been a greater number of different types of vehicles than types of aircraft in service.

Transport vehicles in the RAF for goods and passengers, and certain specialist vehicles, are known collectively as 'MT', which stands for Mechanical Transport, *not* Motor Transport, an important point in the past when RAF transport included engines such as Foden steam wagons, which were mechanical, but not motorised. In this connection, railway engines, both steam and electric, have an association with the RAF, who have operated their own railways.

To trace the full history of the transport of the Royal Air Force, formed April 1 1918 by the amalgamation of the Royal Flying Corps (RFC) and Royal Naval Air Service (RNAS), it is necessary to review the MT of these two Services, as the vehicles they acquired continued in RAF service for many years. So the book opens with details of the transport of these pre-RAF air services.

The RNAS was using armoured cars from the outset of the First World War and when the Army in 1922 handed over control of the policing of Iraq to the RAF, armoured cars were introduced into that service. The RAF Regiment today still operates armoured fighting vehicles (AFVs). So this is a book of AFVs as well as transport and specialist vehicles.

Like aircraft, service vehicles have had their individual identity markings. They were, and still are, subject to decrees on camouflage and markings, with special finishing schemes being appropriate to certain classes of vehicles. Again, as in the case of aircraft, headquarters, stations and units each have an authorised holding of MT, known as their 'establishment', which will vary

according to the function of a unit. The actual number of vehicles operated is known as the 'MT holding'.

A vehicle with an engine to move itself is know as a 'prime mover'; an important distinction, since these are subject to certain licensing regulations which are not applicable to trailers. Towed vehicles are classed as trailers, except in the case of low weapon carriers, which are classed as trolleys.

In the Services, types of MT, unlike aircraft, are classed by their maker's name and designation, or popular name, with a brief indication of function, eg, 'Humber Snipe Saloon' gives the maker, the popular name of a particular model and its function in three words. While this is in popular usage, the official stores designation is a different matter. For example, an AEC Mammoth 10-ton fuel tanker is classed as 'Truck, Tanker, Fuel, 10-ton, 6 × 4, 4,000 galls, AEC Mammoth'. Lorries were called 'tenders', but, with the present inter-Service rationalisation of MT, a unified nomenclature is used and the RAF now refers to 'trucks', 'vans' and 'cars' for general vehicles, then their function, weight and drive, capacity and maker's type.

To explain the 'drive' to the uninitiated—the '6 × 4' given above indicates that the vehicle has six wheels, of which the engine powers four. The first number is always the total number of wheels, and the number after the multiply sign is the number driven by the engine. For example, in the average family car, the engine drives the two back wheels, but not the front two, hence it is a 4 × 2 car. However, a Land-Rover, with power to all four wheels, is a 4 × 4 vehicle.

In the past, horse-power has at times come into the official description of models, and, in certain instances, mark numbers are used to distinguish variations in a particular model. In other cases the manufacturer's model numbers are used. There is no set system.

The book has been divided into two parts. Part One is a chronological survey, whilst Part Two deals with those specialist vehicles, like motor cycles and ambulances, which have carried special markings over their 70 years' use by the air services.

While it is now government policy to use metric in place of Imperial measures, it has been necessary to use the measurements of the day, since they were often incorporated in designations. For example, vehicles were classed by weight under the old Imperial measures, 3-ton and 15 cwt meaning the weight of goods that they could carry, not the weight of the vehicle. Service vehicles are now classed, under NATO agreements, by the tonne; the metric tonne of 1,000 kg which is the equivalent of 2,200 lb. The old Imperial ton was, of course, 2,240 lb.

The idea for a book on RAF MT originated with Bruce Quarrie of Patrick Stephens Limited, when *Airfix Magazine* was under their control and I was asked to write a series under the title 'Wheels of the RAF'. For that magazine, there was concentration on MT markings for the modeller and while these details are included here, the text has been completely re-written and the scope greatly broadened to give a general picture of how the RAF has been served by vehicles over the years. This book is not in any way a reprint of the magazine articles and many new illustrations are presented in this review.

It has been assumed that with so little published on RAF vehicles, those interested may well have *Airfix Magazine Guide 27: Modelling RAF Vehicles* by Gerald Scarborough, published by Patrick Stephens Limited in 1978. For this reason, duplication of any of the collection of illustrations used in that fine little book has been studiously avoided.

Part 1 Chronological Survey

Scene during the 70s; an RAF Bedford truck-mounted extending service platform is used for a check on the fin of a visiting USAF B-52F of the 93rd Bomb Wing from Castle Air Force Base, California (USAF).

Chapter 1

Royal Flying Corps MT

The Royal Flying Corps (RFC) was formed on May 13 1912 from the Air Battalion, Royal Engineers, as an integral part of the British Army. The Army had been using limited MT, particularly steam traction engines, since before the turn of this century for heavy gun haulage. In 1911 a decision was taken to replace horse transport by motor transport on a large scale to achieve greater mobility, but finance was the problem. The Treasury sanctioned only a limited change-over, insufficient to meet the War Office plans for any future war.

The new RFC was high on the list for mechanical transport as a result of the 1912 Army manoeuvres in which it played a significant part. An analysis showed that it was essential to have rapid communication between aircraft crews and troops and that landing grounds would have to be established near the headquarters of Army Commanders and even Divisional Commanders. The stagnation of trench warfare was then undreamed of and mobility was seen as of the utmost importance. It was realised that the new arm would involve a large ground organisation to fulfil its commitments, for in the manoeuvres eight steam wagons, ten heavy lorries, twelve light tenders and eight motor cars had been needed from the transport pool to keep two airships and fourteen aeroplanes in commission. From War Office vehicle trials that year in the 3-ton and 15 cwt classes, the RFC was allotted Leyland lorries and Crossley tenders and touring cars as their standard squadron equipment.

To war on wheels

Even at this early stage it will be seen that to operate expediently, an air service would need more wheels than wings. But when war came and the four squadrons nominated for the British Expeditionary Force mobilised at Farnborough, their establishment of MT was inadequate. Fortunately there was a ready supply from the London streets.

Knowing the limitations imposed by the Treasury for peacetime scales of equipment, the War Office had the foresight to institute a subsidy scheme following their vehicle evaluation trials. Firms with approved vehicles in the 3-ton and 15 cwt class were assisted in a modest way for the purchase of their MT, £50 towards purchase and £20 per annum for maintaining it to precise standards, and on the understanding that in an emergency the Army would purchase it for full-time use. When war was declared in August 1914, over 1,000 vehicles in 3-ton and 15 cwt categories had been registered under the scheme. Leylands allotted to the RFC were collected at Regent's Park, then a party of RFC drivers from Farnborough drove them to Southampton for shipment as they were.

Early type Leyland issued to the Royal Flying Corps on its formation, registered AA5036 (MoD H1907).

One of the lorries allotted to No 5 Squadron from the London store of Maple's was of a brilliant scarlet hue with an advert from a famous sauce manufacturer proclaiming it 'The World's Appetizer'. Used as the squadron's ammunition carrier, it served as a landmark for returning pilots. At a later stage of the war it would have invited bombing attack.

As well as their nomal transport and subsidy scheme allotments, there were other additions, particularly cars, impressed for war service or donated by patriotic civilians. No 5 Squadron had Mercedes, Renault and Sunbeam touring cars on strength, plus an attached civilian two-seat coupé complete with owner, recorded as a Mr Weeding of Guildford.

The motor industry's effort

Following the declaration of war, the War Office ordered transport on a large scale. The basic heavy transport vehicle of the British Army was the 3-ton lorry and for these, contracts were placed with Albion, British Berna, Daimler, Dennis, Karrier, Maudsley, Pagefield and Thornycroft. Additionally Halley, Halford, Napier, Wolseley, Lacre and Straker-Squire lorries were used in more limited numbers and mainly in the United Kingdom. Examples of most of these types found their way into RFC units, but Leyland was an exception, for their output of 3-tonners was reserved almost exclusively for the RFC.

For 30 cwt vans and tenders the War Office made acquisitions from Belsize, Daimler, Napier and Straker-Squire for general use, with the entire output of Crossley tenders going to the RFC. Later vans were purchased in America from Ford, Garford, Studebaker and Willys Overland. The standard Army ambulance was a Sunbeam, but others were acquired from Austin, Daimler, Napier, Siddeley Deasy, Wolseley and later, in particular for RFC airfields in the UK, Fords.

British Army staff cars were of three standard types; Sunbeam (built by Rover post-1916), Vauxhall for general use and Crossley touring cars for the RFC in particular. Limited numbers of Austin, Daimler, Rolls-Royce, Singer

Crossley tender salvaging a No 6 Squadron BE2 in Flanders 1914.

and Wolseley cars were also in use. Since Crossley staff cars were issued down to squadron level, the more senior RFC officers at wing and brigade level tended to prefer other types.

After 1915, Daimler, as the first motor firm to build both aero engines and airframes, gave priority to aircraft production and passed over their motor transport contracts to the Associated Equipment Company (AEC) of Southall, Middlesex. Similarly Sunbeam, building engines for airships and seaplanes, both on Admiralty contracts, passed all their automobile work in 1916 to The Rover Company, which produced vehicles at their Solihull works to Sunbeam designs.

Vehicle establishments and strengths

By 1916, with the motor industry geared to war tempo production, the establishment scales were soon met. For the RFC these were on a lavish scale. Each squadron was allowed eight Leyland heavy tenders, four Leyland workshop lorries, eleven Crossley light tenders, or Fords or other types in lieu, two cars (normally Crossleys), four trailers, four motor cycles with sidecars and four without. An Army Corps squadron had a more lavish allowance of motor cycles for despatch riders.

The 'CO's car' was a term loosely used, but the Commanding Officer of a squadron, at the level of Major, did not qualify for a car to be allotted personally. Naturally as commander of a squadron, normally with two cars, he would have first call on his own unit's transport and in effect annex one car for his exclusive use. But by establishment, the officer commanding a wing, at Lieutenant-Colonel ranking, was the lowest level to which a car was issued personally.

The scales were laid down for a war of mobility. When there was trench warfare, transport was not fully utilised in the static airfield conditions, except for the workshop lorries. Nevertheless, the vehicles had to be kept on stand-by for offensives by either side, in case of advance or withdrawal. In any move the loads were quite considerable in tentage alone. There were the large Bessoneaux hangars, spares including aero engines, technical and accommodation stores

A Subsidy Scheme Leyland sketched from a photograph of this vehicle standing by the Cloth Hall, Ypres (Peter G. Cooksley).

and arms and ammunition, all to be put on the road. The RFC could not rely on the motor companies of the Army Service Corps, which would have had tasks enough in any forward or backward movement of the front.

The need to conserve transport, yet to have it available for an emergency led to a new RFC holding policy in the field from September 1917. Surplus heavy tenders were withdrawn from squadrons and RFC brigade lorry parks were formed at each of the six RFC brigades in France and Belgium with establishment for 30 Leylands and 24 trailers. This was later increased by 15 lorries and 12 trailers just in time for the great March 1918 German offensive, when many squadrons, plus a complete air depot, had to pull back. It should be appreciated that to uproot a single standard Bessoneaux hangar and transport it to a new location for re-erection involved three 3-ton tenders and trailers to accommodate the canvas and the framework, and a handling party of around 20 men.

The numbers of vehicles overall on RFC charge rose steadily. In May 1915, when the RFC operated 156 aircraft on the Western Front, they were served by 351 vehicles. At home, at that time, 370 vehicles were utilised to serve 250 aircraft, and orders had been placed for 640 more vehicles. Vehicle strength then rose swiftly as follows:

Date	*Home*	*Overseas*
June 12 1916	1,445	1,921
September 29 1916	2,804	2,267
March 1 1917	3,518	3,467

By the time the RFC was integrated into the RAF on April 1 1918, the Corps had over 6,000 tenders and cars and some 2,500 motor cycles, mainly P&M type.

MT field supply

In France the Army's mechanical transport was serviced by mobile Army Service Corps (ASC) workshops, but the RFC, having its own workshops, undertook minor repairs. Major repairs throughout the Army were the

An RFC Leyland shows its capacity for carrying a complete aircraft, fuselage in the centre with wings lashed to the sides (J.M. Bruce/G.S. Leslie collection).

responsibility of the Army Ordnance Department, but with some 20,000 lorries alone in use by 1916, Ordnance, which was responsible also for artillery and limber repairs, could not cope and the ASC assumed responsibility for repairing all Army MT—including vehicles of the RFC—and opened three large MT Base Repair Depots.

The RFC sent its tenders to the depots at Rouen or Paris and its bicycles to Rouen, while its tourers went to the Le Havre depot, which dealt with ambulances and cars. New vehicles were issued by the Base MT Depot, ASC, at Abbeville. This base was subsequently moved back to Calais because of German air raids. When the RAF was formed, a large new MT depot was built at Motteville, and this gradually assumed responsibility for all RAF transport. This proved fortunate for the RAF, for the great Calais depot was bombed on August 11 1918, starting a conflagration at the depot that could not be brought under control for hours. Spare parts for 19,000 lorries and cars were lost, together with the vehicle ledger cards and records of 14 makes of vehicles. Records for Leylands and Crossleys had by then been transferred to two separate office blocks at Motteville.

Finish and identity numbers

Like Army vehicles, those of the RFC were in standard khaki finish, except for touring cars, which often retained their commercial finishes at the discretion of commanding officers. In such cases the finish was not even compromised by identity numbers, apart from civil registrations, as these were marked on the dashboard. There was no standard RFC marking. The khaki implied Army, though the Leyland, Crossley and P&M types of lorries, tenders and motor

An all too familiar sight, RFC Crossley tender broken down by the wayside (Capt D.S. Glover).

Leyland chart lorry of the RFC outside a chateâu taken over as a headquarters (Cosford A0598/1).

cycles were almost exclusively RFC. In the Expeditionary Forces, as a security measure, unit markings were not applied, as it was thought that enemy agents were posing as refugees behind the fronts. At home some training squadrons and acceptance parks marked their cab doors with unit initials, but there was no known official edict on this matter.

At first, individual vehicles were recorded both by their maker's engine number and by their registration number, allotted by the local authority in the United Kingdom. Since the registration number had little significance overseas, and there was little point in licensing vehicles going to units mobilising for overseas, a Service-issued registration number was marked in white or chrome yellow on the vehicle, with a 'C' suffix to denote the series. Examples of the numbers in the 'C' series are given here from vehicles allotted to the famous No 56 Squadron at London Colney when working up for overseas service in the summer of 1917: Crossley touring car No 7484, registered 34585C; Crossley light tenders Nos 7536-7538, registrations 34633-34636C; and Leyland 3-ton lorries registered 29929C, 35210C, 37108C and 38204C.

Chapter 2

Royal Naval Air Service MT

The Navy's air arm had started as the Naval Wing, Royal Flying Corps, with a small establishment of Royal Navy vehicles supplied for general transport. From July 1 1914, the Naval Wing became a purely naval service—the Royal Naval Air Service—and its allotted vehicles, previously marked 'RN', were marked 'RNAS'. When war came, there was a rapid expansion and the transport element was swelled by presented and impressed vehicles and large orders were placed with the motor industry.

RNAS transport vehicles

The Admiralty had a close association with the Clement Talbot works in Shepherd's Bush, London. The famous French Clement car had been introduced into England through the financial backing of the Earl of Shrewsbury and a man named Talbot, hence the name. The fact that the company was a French concern proved very helpful when the RNAS was based in France and required spares urgently. The RNAS standard tender, equating to the RFC's Crossley, was the Clement Talbot 15 cwt 4 × 2 tender. The Admiralty acquired over 1,000 vehicles from Clement Talbot Ltd, many for allocation to the RNAS as general utility vehicles. The firm had its resources stretched, being also used by the Admiralty for aero engine and limited airframe repair work. Light tenders were also acquired from the Lancaster Motor Company of Birmingham, which, from late 1915, built aero engines.

RNAS vehicles were numbered in a simple numerical series, starting at No 1, prefixed by 'S' to denote the series. Vying with the Army to place contracts for vehicles in the UK, the Navy, with purchasing officers in America buying Curtiss aircraft, also placed orders there for vehicles, both armoured and transport. The drawback to this was the great variety of spares, and delivery difficulties.

RNAS armour

The concept of an armoured aircraft recovery service, operating armoured vehicles, was rapidly developed into an armoured car division of the RNAS in 1914. Making Dunkirk their base, the first RNAS armoured units in France were given the task of ensuring that the Germans did not venture within 50 miles of the base. For this purpose the RNAS was informed by the Admiralty that it would be reinforced by 200–300 officers and men equipped with 50–60 armed motor cars, to support the planned three squadrons of 12 aircraft each which would comprise the 1st Wing.

Clement Talbot 15 cwt standard RNAS light tender, RNAS S137, with the letters RNAS originally in white on the grey naval finish practically painted out by a khaki wash over the side-boards before shipment overseas (L.A. Dell).

To support the armed and armoured cars, additional vehicles were required as supply and ammunition carriers, breakdown vehicles and ambulances, as well as touring cars and motor cycles for liaison work. Eventually over 200 vehicles were listed on RNAS strength in France and Belgium as follows: 35 armoured Rolls-Royces of which 15 were armed with twin Maxim machine-guns, 42 armoured Wolseleys of which 15 had twin Maxim machine-guns, 8 each Wolseley breakdown trucks and ambulances, 22 Wolseley transport wagons, 46 armoured Talbots (15 fitted with twin Maxim machine-guns), 18 Talbot wagons, 8 each Talbot breakdown trucks and touring cars and 10 motor cycles of type unknown.

In mid-September 1914, when the front was still fluid, the armoured cars proved their worth, attacking parties of Uhlans scouting forward of the main body of the German Army. Commander G.R. Samson in charge of operations from Dunkirk was in the forefront of the action. He had first gone into action against the Germans in a touring car which had been fitted with a machine-gun. After recall to England he organised the motorised formation which officially became the Aeroplane Armoured Motor Support, RNAS, organised into four squadrons of four sections each.

The Dunkirk depot was expanded to process new vehicles with the requisite armour, carry out repairs on those in service and conduct experimental work. One 30 hp Wolseley car, which had its back axle break under the weight of its armour, led to Wolseleys being withdrawn. On the other hand, the personal Mercedes of Lieutenant F.R. Samson could still touch 55 mph with full armour. The armoured personnel carrier was heralded by two ex-London General Omnibus Company lorries, protected by all-round armour, with loopholes as firing ports for riflemen.

Commander Samson moved his mobile force to a new headquarters at Hazebrouck on September 19 1914, leaving the depot the task of standardising for a force of 50 Rolls-Royce armoured cars, fitted with Lewis guns in place of the Maxims. The 6 mm steel plate used as armour was supplied by the *Forges et Chantiers de France* plant at Dunkirk. The varied assortment of vehicles was to

A commandeered Rolls-Royce Silver Ghost, seen in London 1915, still wearing its civil registration LM-3102 (L.A. Dell).

be replaced by the proven Mercedes, using the Maxim guns taken from the other vehicles, which were mounted to fire forward for ground operations, and be elevated when in an anti-aircraft role. As it was, with trench warfare starting, German airships abandoning day reconnaissances over land, and German aeroplanes being rarely seen at that stage of the war, the vehicles were under-employed and were attached to Army units for ground support work.

Further detachments were sent to the continent. Headed by Major Risk, Royal Marines, who had his own touring car with him, six armoured cars and two wagons left Sheerness for Dunkirk on September 25 and were sent immediately to Morbecque on patrol. At the end of the month, this new force was withdrawn to the Dunkirk depot to join five more armoured cars, some lorries and an ambulance, sent over on the seaplane carrier HMS *Empress.* They were then ordered to escort 16 buses carrying stores for the naval landing at Antwerp, which was switched to Ostend when the former city fell. The Hippodrome at Ostend was taken over as an MT depot, but, like Antwerp, the town had to be evacuated quickly in the face of the German advance.

So far the armoured cars had carried nothing heavier than machine-guns. Under the supervision of Lieutenant Warner RM at the Dunkirk shipbuilding yards, a 3 pdr quick-firing Hotchkiss gun was mounted on a Milnes-Daimler lorry. On October 19 this prototype self-propelled gun went into action firing 80 rounds that day and continued in operational use until the gun rifling wore smooth. Further lorries were similarly modified.

November weather and heavy traffic caused the condition of roads near the front to deteriorate and the movement of heavy vehicles was severely hampered. Because of this, Commander Samson asked for the next 3 pdr Hotchkiss to be mounted on two wheels, to permit it to be towed behind the touring car as a trailer. The French requested RNAS armoured car support for their attack along the Yser that month, but the winter and trench warfare put a stop to further mobility.

Left *The transport backing for the armoured car squadrons going to the Dardanelles, seen after collection outside the Clement Talbot works* (L.A. Dell).

Centre left *Lanchesters of A Section No 15 Squadron before leaving for Russia where the Section became No 1 Squadron of British Naval Armoured Cars, Russia* (L.A. Dell).

Bottom left *A section of anti-aircraft Seabrooks, with motor cycles and armed guard, halted in an East Anglian town in 1915* (L.A. Dell).

Top right *Peerless 3-ton tender of RNAS Vendôme in France. Being far from the Front, at a flying training station, the lorry is painted in navy blue and, like the RFC/RAF Leylands, carries a loading warning* (B.F. Hansley).

Centre right *The proven armoured car of the RNAS that went on to do yeoman service in the RAF, the Rolls-Royce, here seen with its RNAS crew* (L.A. Dell).

Right *The haulage and ammunition wagons of the armoured car squadrons were Jeffrey Quads, with four-wheel drive and two-axle steering. This vehicle was allotted, as the letters 11 B4 RNAS faintly discernible on the original negative show, to B Section, No 11 Squadron, RNAS* (L.A. Dell).

LOA NOT
TO EXCEED
TONS
FRED. KARNO'S
AIRMEN.

RNAS armoured support was re-organised in February 1915. There was call for armour in other theatres of war and in France there was optimism that in the spring a new large-scale offensive would lead to a resumption of open warfare. Meanwhile Commander Samson had been sent to the Aegean island of Mudros to organise air support for the Dardanelles campaign.

Armour in Africa

No 1 Squadron, based on the Dunkirk depot, was withdrawn to Britain for refitting to assist South African troops in the conquest of German South-West Africa. Arriving at Walvis Bay April 17 1915, their 12 vehicles, mainly Rolls-Royce armoured cars, went by train to Trekkopje. When the camp there was attacked nine days later, the cars played an important part in repulsing the enemy, coming as they did as a complete surprise to the German-officered native troops. This was in spite of their having been observed by a German reconnaissance aircraft whose pilot, unused to such sightings, had reported the cars as field kitchens! Several cars were hit by shrapnel, but none was put out of action and one man wounded was their only casualty. Resistance came to an end in July, whereupon the squadron was shipped back to England, with the exception of four armoured cars and their crews, under Lieutenant H.G. Nalder, which were sent across the continent to the forces engaged in the conquest of German East Africa.

In East Africa, the armoured cars worked at first in conjunction with No 26 Squadron RFC newly arrived from England. This was a very conventional RFC squadron, having standardised equipment—BE2C biplanes, Leyland heavy tenders and workshop lorries and Crossley light tenders—while the armoured cars, forming a new No 7 Squadron, were given Ford vans and a workshop lorry for support. In late 1916, Howard Nalder, then a Lieutenant Commander, was gravely concerned with the poor health of his force, which was debilitated with tropical diseases, and he received permission to withdraw. The personnel and their cars sailed from Kilindi for Port Sudan and the cars were eventually taken over by the Army in Egypt.

Armour in the Aegean

It was hoped early in 1915 that Commander Samson's earlier successes in Belgium against the Germans could be repeated against the Turks in the Dardanelles Campaign. Nos 3 and 4 Squadrons of armoured cars, and No 10 Squadron with Scott motor cycles armed with Maxim guns on sidecars were shipped to Mudros for the Gallipoli landings. The Maxim guns of No 3 Squadron, embarked on the SS *River Clyde*, from which troops were mown down by Turkish fire as they landed, went some way in preventing higher casualties, but landing vehicles at this stage was out of the question. No 4 Squadron disembarked at Skyros and went later to Gallipoli to support the infantry with their machine guns dismounted. Similarly No 10's Maxims were used in a dismounted role.

The first armoured cars on Gallipoli were reported to be four Rolls-Royces which arrived on May 26 1915, followed by four more on June 3, ferried in on lighters for an attack the following day. This entailed a prior night reconnaissance along the only two roads and a track suitable for use, and with such lack of ability to manoeuvre, the cars soon became casualties. Nos 9 and 10 Squadrons sent out from Britain were held for the Sulva landings, at another

Close-up of a Seabrook of B Section, No 14 Squadron, as its markings indicate. These vehicles had left-hand drive, sliding turret and a crew of five to drive and man the 3 pdr Hotchkiss gun and a Maxim machine gun (L.A. Dell).

part of Gallipoli, but they too acted in a dismounted role and No 12, without cars at all, operated as three machine-gun detachments. No more armoured car operations were attempted.

Peak and end of RNAS armour

No 2 Squadron, remaining in France in the Ypres area, was brought into action on May 13 1915. Three cars of 'B' Section, attached to the Army's 3rd Cavalry Division, participated in a counter-attack to recover lost ground. Although one car was put out of action by shrapnel, the other two cars, *Bustler* and *Busy*, negotiated the shell-pocked roads, forcing the enemy back. That day *Busy*'s crew expended 6,500 rounds from their Maxim machine-gun, most of them directed at close range on the retreating enemy.

By this time the hasty conversions of 1914 had been completely replaced with standardisation on Lanchester and particularly Rolls-Royce armoured cars, armed with .303 calibre machine-guns. These were mainly Maxims, as the Army had first call on the lighter Vickers version and Lewis guns which (while not in general Navy use at that time) were officially allotted to the RNAS. Cars so armed and armoured were classed as light armoured cars and were attached to cavalry formations which rarely again had the opportunity to operate, as the much sought 'break-through' did not come until late 1918. So they were placed in reserve, except for occasional use in delivering despatches.

The earlier lorries with their mounted Hotchkiss guns had been replaced by Seabrook heavy lorries, purchased in America and armoured by the Portholme Aerodrome Company. These large metal-clads gave better protection to the crew manning their 3 pdr Hotchkiss or 1 pdr Vickers guns than the former lorries. They were deployed by GHQ, BEF, at points along the front, as a mobile artillery reserve, taking up position in the dark, opening fire at dawn, and withdrawing before counter-battery fire endangered their position. The Army, however, while appreciating their mobility, was not impressed with the relatively small calibre of the guns and doubted their accuracy without the normal battery work of forward observation post information, prior zeroing and ranging. There was also the problem of communications with adjacent field formations. While the Seabrooks' fire would be invaluable in an offensive, their guns were ineffectual in trench warfare. The hopes of using them to bring some form of artillery to the forefront of any advance soon faded with the realisation that, with the continuous artillery pounding, the ground between and on either side of the lines would be impassable to wheeled vehicles.

Altogether, in France in the spring of 1915, there were 36 light armoured cars and 18 heavy armoured lorries. They were deployed in five squadrons, Nos 2, 6 and 15 allotted in sections along the front and Nos 5 and 8 based on Dunkirk. But in August 1915, all these armoured units were transferred to the Army.

Home Front armour

The RNAS played a part in the ground air defences of the UK when such defence was an Admiralty responsibility in the opening years of the war. While a Royal Navy Mobile Anti-Aircraft Brigade was allocated to the defence of London, there were two mobile anti-Zeppelin RNAS armoured car squadrons with the task of catching the Zeppelins between the coast and London. These two squadrons were Nos 14 and 18, each operating three sections of three Seabrook heavy armoured cars, mounting a 3 pdr Hotchkiss and .303 Maxim machine-guns. The Seabrooks were prepared at the Clement Talbot works on receipt from America. They operated in sections on the East Coast from Northumberland to Essex. With a speed about half that of a Zeppelin, it was thought that they would have a radius of action to interpose themselves between the coast and the target on the Zeppelins' homeward route. As they thundered along country roads to take up position following attack reports, it was at times feared that they were more of a menace to country folk than the Zeppelins.

When in August 1915 the Army took over control of these armoured cars from the Admiralty, No 20 Squadron remained under the RNAS. They tested a fore-runner of the tank at their Wormwood Scrubs base. This was the Killen Strait tractor which was fitted with a turret taken from an armoured car. In

A training Seabrook (R1 for Reserve No 1), fitted with a mast for flying the White Ensign as a landship (L.A. Dell).

mid-1915 the squadron moved to the Trench Warfare Department's experimental ground at Wembley Park to further their experiments. Some 30 officers and 200 rating were engaged on this work in 1918, under Commander R.W. McGrath RNVR.

RNAS armoured car squadrons, 1915

Squadron	*Main equipment*	*Area of operations*
1	Rolls-Royce	South-West Africa
2	Rolls-Royce	Belgium and France
3	Rolls-Royce	Dardanelles (dismounted)
4	Rolls-Royce	Dardanelles
5	Lanchester and Seabrook	Dunkirk-based
6	Lanchester and Seabrook	Belgium and France
7	Rolls-Royce and Ford	East Africa (4 cars ex-No 1)
8	Rolls-Royce and Seabrook	Dunkirk-based
9	Various	Dardanelles (dismounted)
10	Motor cycle and sidecar	Dardanelles (dismounted)
11	Motor cycle and sidecar	Dardanelles (dismounted)
12	None	Dardanelles
13	Motor cycle and sidecar	Not fully mobilised
14	Various	East Anglia
15	Rolls-Royce and Lanchester	Belgium and France
16	Seabrook only	Belgium and France
17	Seabrook only	Belgium and France
18	Seabrook	East Anglia
19	Seabrook	Not fully mobilised
20	Various experimental	United Kingdom

Squadrons in France in the main were divided into three sections of four light cars plus a section of three heavy cars, but the heavy Seabrooks were later withdrawn to form Nos 16 and 17 Squadrons.

RNAS to RAF

One operational squadron remained under RNAS control, this was No 15, sent to Russia as the British Armoured Car Division. Reorganised into Nos 1-3 Squadrons it operated Lanchesters and Seabrooks, with a single Rolls-Royce, plus transport including motor cycles. After supporting Russian forces against the Turks in the Caucasus, the squadrons were switched in late 1916 to Romania against Bulgarian and German troops. They received reinforcement and replacement vehicles from the Russians, including some of British make. After the Russian Revolution the personnel withdrew to England leaving their vehicles in Soviet hands.

This was the end of naval armoured units operated by the RNAS, which, nevertheless, continued to expand and require trucks and tenders for its coastal duties around the British Isles, its training stations and depots, and its growing bombing force. However, for its detached fighter squadrons operating with the RFC, Army transport was provided to supplement the RNAS tenders which plied between the Dunkirk Depot and the detached squadrons. With the formation of the Ministry of Munitions in 1916, followed by the creation of an Air Ministry in 1917 and then the amalgamation with the RFC as the RAF, there was progressive standardisation of air arms throughout the First World War.

Chapter 3

MT of the Commonwealth Air Forces 1915-19

While thousands of Commonwealth personnel served in the RFC, RNAS and RAF, both Australia and Canada raised air arms of their own, using British aircraft and vehicles. Additionally, a Training Brigade of the RFC was raised in Canada to train recruits from the American continent, using Canadian-built Curtiss JN4s and local transport. 'Local' in this context included the hub of the world's motor industry in the adjoining United States.

Australian Flying Corps

The Australian Flying Corps had its beginnings in a half-flight, formed with Australian personnel, for service in Mesopotamia in conjunction with the Indian Government. Initial agreements were that the unit's transport would be both animal (horse and mule) and mechanical. Two workshop lorries were supplied by Australia, built by the Newport Railway Workshops, equipped with benches, lathes, drilling machines, circular saws and a welding plant. To these were added three Berliet trucks, a staff car and an undisclosed vehicle. Numbered AFC 1 to 7, these vehicles reached Basra via India and were used by AFC, RFC and RNAS personnel in Mesopotamia.

The first AFC squadron, No 67 RFC (later re-numbered No 1 AFC), took up station at Heliopolis in Egypt, where aircraft and transport were supplied from RFC resources and used later in the squadron's move to Palestine. On the Western Front, Nos 68 (DH5s later SE5As), 69 (RE8s) and 71 (Camels later Snipes) Australian Squadrons, RFC, became Nos 2, 3 and 4 Squadrons AFC. As they were brigaded with RFC formations, standard Leylands and Crossleys were supplied from RFC stocks in the field.

To provide a trained backing for the Australian squadrons on the Western Front, the 1st Australian Training Wing was set up in England with headquarters at Tetbury, Gloucestershire, with Nos 5 and 6 Training Squadrons at Minchinhampton and Nos 7 and 8 at Leighterton. There was also an AFC Depot at Halesford Camp and a Trades Training School at Halton. To each of these establishments MT was provided from RFC resources, as were their training aircraft.

In the United Kingdom, since the Australian Government contributed funds to finance the AFC units, their MT was numbered in a special series from 1 upwards, qualified by the letter A for Australia applied as a suffix to avoid confusion with local authority allocations. Examples are 4A Royal Enfield motor cycle and sidecar, 108A Minerva touring car and 312A Ford model 8T van. The AFC was disbanded in 1919.

Above *Crossley tender at the AFC Station Minchinhampton towing Avro 504K fuselage from a Bessoneaux hangar of No 6 Squadron AFC, 1918* (Australian War Memorial E5530).

Below *Ford fire tender at Camp Leaside RAF (Canada) training station, 1918. The red Ford, bearing the station location of North Toronto on its bonnet, is shackled to a fire equipment trailer; an alternative ladder trailer is seen adjacent to the lean-to on the left* (Public Archives of Canada PA22909).

A Packard tender waits outside the Orderly Room of Camp Mohawk, Deseronto, Ontario, in 1918 (Public Archives of Canada C83061).

RFC/RAF Canada

To make use of the vast flat tracts of country far removed from the theatres of war, and to train within the North American continent the growing number of RFC recruits from Canada and the United States, an RFC Training Brigade was established in Canada early in 1917. To provide MT for the Brigade Headquarters and training stations, a Mechanical Transport Section was formed at Toronto in March 1917. It was housed in the Wolseley garage initially, until it moved to a large building in Dupont Street. Shortly before the war ended, even larger premises were acquired on Avenue Road for garaging, leaving the former premises for repair work.

Local transport was used, with the exception of two Leyland workshop lorries shipped out from England for use at Camp Borden before workshop facilities were built. After initial local acquisition of staff cars, purchases were made standardising on types of specialist vehicles. The Toronto MT Depot employed over 250 staff, including lady drivers, and operated some 400 vehicles. This was apart from vehicles issued to flying stations, and their drivers.

Commercial finishes were allowed on staff cars but, while camouflage was not necessary, other vehicles were finished in khaki for a uniformity which was appropriate to Army vehicles. When the RAF was formed on April 1 1918, it was only a change on paper from RFC (Canada) to RAF (Canada). The exceptions to the finishing schemes were, appropriately, the fire engines in bright red and ambulances in white.

Vehicles were registered in blocks of numbers allotted by arrangement with provincial authorities. These numbers were normally presented front and rear of the vehicles with a 'C' prefix to denote the series. Examples are as tabled. Trailers were numbered from C101 to C150 and Indian side-cars were not given a number other than their manufacturers number.

Representative vehicles of RAF (Canada), 1918

Type by official description	*Chassis No*	*Service No*	*Unit*
Ford general utility truck	165543	C4045	Depot
	156631	C4084	Not known
Ford gasoline tanker truck	148051	C4021	93 Squadron
	148094	C4026	Depot
Ford fire engine	2477900	C4080	Depot
Indian motor cycle	82J028	C133	Depot
	82J015	C141	Depot
Kelly Springfield roller	3418	C1	School of Aerial Fighting
Packard float lorry	105254	C1132	Depot
	107277	C4578	Depot
Packard heavy tender	75659	C1106	School of Aerial Fighting
	106134	C1112	School of Aerial Fighting
Packard gasoline tanker	93	C1094	Depot
Packard utility truck	75656	C1098	Depot
	75657	C1103	Depot
Packard ambulance	132456	C1140	Not known
	131728	C4078	School of Aerial Fighting
Studebaker touring car	221705	C1024	93 Squadron
	221592	C1025	93 Squadron
Studebaker combination	231640	C4057	Not known
	231012	C4069	Not known
Studebaker tender	210855	C1045	Depot
	216567	C1081	Not known

RAF Leyland 3-ton tenders and workshop lorries of Nos 1 and 2 Squadrons, Canadian Air Force, in a hangar at Upper Heyford, Oxfordshire, early 1919. An Indian-head marking appeared on cab doors (E.F. Cheesman).

Packard, the standard ambulance for RFC/RAF (Canada) 1917-18 (RCAF RE19570).

Canadian Air Force

A study of Royal Canadian Air Force MT is outside the scope of this book, as is the MT of other Commonwealth Air Forces. The original CAF was formed in Britain by Nos 93 and 123 Squadrons RAF being nominated Canadian squadrons, to be equipped with Sopwith Dolphins and DH9s respectively. These were supplied from RAF sources, together with the supporting MT. No 93 Squadron was re-designated No 81 Squadron on October 19 1918. These two RAF squadrons were re-named Nos 1 and 2 Squadrons CAF and were mobilised at Upper Heyford shortly after the Armistice in November 1918. Their MT had their first convoy experience at the end of March 1919 when the squadron moved to Shoreham as the Canadian Wing, shortly before the force disbanded.

During 1919, Commwealth countries were offered an Imperial Gift of 100 aircraft each, plus ancillary equipment, to form their own air forces. In the case of Canada, the equipment included 300 vehicles, yet again emphasizing that an air force needs more wheels than wings. The air arms of Australia, Canada, New Zealand and South Africa initially modelled their MT On the RAF, acquiring Leyland workshop lorries and tenders and Crossley light tenders. In the 30s these forces provisioned their own MT, which comes outside the scope of the RAF and therefore this book, but where Commonwealth squadrons operated under RAF command, in the Second World War, they received their MT from the RAF.

Chapter 4

MT in the RAF's formative years

King's Regulations for the new Royal Air Force were issued on March 26 1918, five days before the new force came into being. They contained detailed instructions on many aspects of the new Service, including the marking of vehicles with the initials RAF. However, like many other matters of service policy, until the war ended, uniforms, customs, accounting and aspects like vehicle markings continued as before; so that the RAF throughout 1918 operated vehicles marked WD, RNAS and RAF.

There are many stories of airmen driving RAF tenders abroad at hair-raising speeds when they were used for off-duty runs to local towns, and certainly there were a number of casualties sustained in this way. But in the UK it was different. In mid-1918 the Air Ministry issued the following order to apply to units at home: 'Great damage is being done to roads in consequence of the excessive speed at which Motor Transport Vehicles are being driven. The following speeds: 2- and 3-ton lorries, 8 miles per hour; 30-cwt and lesser lorries, 12 miles per hour; should never be exceeded, save in exceptional circumstances. If absolutely necessary for urgent service reasons the Commanding Officer may give permission for the above speeds to be exceeded'.

Run-down and reconstitution

In late 1918 the RAF had 23,260 MT vehicles on strength, not including trolleys, handcarts and the like. While demobilisation proceeded, vehicles were stored in hangars, while MT in use was repaired at the RAF MT Repair Depot, Harlescott, near Shrewsbury. Bids by private firms were put in to buy up the vast stocks at home and in France. With a lack of shipping, and to save transit costs, MT in all other overseas theatres was disposed of locally.

From France, apart from shipment from the channel ports, some MT came home on barges. To give some idea of the magnitude of the task and the chaos existing because of key personnel being demobilised and staff constantly changing, the state of some of the vehicle types from France is detailed. There were 838 3-ton Leylands handed over to a government depot at Staines, from something like 1,000 which should have been on books; another 3,000 were purchased back by Leyland for reconditioning, mainly for the commercial market. Of 534 Crossley tenders recorded in transit, only 468 arrived. It is evident that there was large-scale pilfering both in France and in Britain. After a Crossley was stolen in London, orders were issued that if left unattended by drivers for duty calls, they were to have the wheels padlocked. On the other hand, of 101 Crossley touring cars expected back, 107 actually arrived! From

A Hendon-based Crossley Type P fire tender of the RAF in bright red finish, seen registered in the same way as civil vehicles in accordance with the regulations of the period (MoD H500).

this it was deduced that there was not so much pilfering by individuals, but organised crime by industrial organisations. A single Jeffery Quad and a Pierce Arrow and a Lacre workshop lorry survived, but the single Lacre 30 cwt lorry listed on books was reported stolen. Hundreds of trailers returned, and as hangarage had been allotted to prime movers, most had to be stored in the open. With so many trailers already in the UK, they created quite a problem of space, for there were 669 wing and 618 flat trailers alone. Many were sold and could be seen gaily painted during the 20s and 30s on fairgrounds. Some vehicles were scrapped out of hand. All 12 travelling kitchens which returned were rusted up and useless.

From this mass of returning transport, plus that in service in the UK, the RAF had to decide on its MT policy as a peacetime force. With the government planning policy that there would be no major war for ten years, so no

Expeditionary Force was envisaged, squadrons would not need mobility on their permanent stations, and in consequence MT was provided on a much more limited scale. While the unit establishment was scaled down to the minimum, depot strength holding was high, to effect replacements and provide spares without buying new equipment. This was at a time when the Geddes axe was poised over the Services and the very necessity of a third Service was being questioned.

The RAF was divided in the early 20s into two main commands; Inland and Coastal Areas. The Inland Area Aircraft Depot at Henlow was also the main storage depot for vehicles, holding over 400, many of which required repair. There was only one RAF MT repair depot at the time, at Harlescott, with a capacity to overhaul 300 vehicles a year. Units at home sent vehicles for repair to Harlescott. Depending on the work-load, vehicles were fed in from Henlow to decrease the repair backlog.

For racing to resume at Ascot in 1919, the RAF had to quickly evacuate the grandstand used as an MT store and distributing centre, but so vast had the depot grown, that areas outside the stands and track were also used, and with the reduced peacetime establishment of vehicles this became the RAF's packing depot. Over 200 vehicles per year were being packed there in the early 20s, as were aircraft, to meet overseas commitments. MT spares were held centrally in London at No 1 Stores Depot, Kidbrooke.

The first of the peacetime establishments for MT is tabled. Of the personnel allotted for MT, the bulk were drivers, with a fifth of the allotment being fitters. With only 32 officers concerned with MT throughout the whole of the RAF, there was the astounding ratio of less than one officer per 100 men employed on MT. However, it should be appreciated that for disciplinary purposes, drivers and fitters at stations would come under non-MT specialist officers. Overseas, included in the MT personnel establishment, was a total of 12 civilians and 125 recorded as 'natives'.

In India, ambulances for the RAF were on Indian Government charge, numbered in their own series prefixed by a broad arrow, supplied to military cantonments with medical Aid Centres (hence No 3MAC) for Army and RAF use. Vehicles used were standard types for the British services, the ambulance shown having a Ford Model T chassis (F.W. Lord via P.H.T. Green).

First RAF peacetime vehicle/personnel establishment, November 1922

Unit/Units	*Cars, various*	*Leyland tenders*	*Crossley tenders*	*Ambulances*	*Miscellaneous*	*Drivers & fitters*	*Remarks*
Inland Area HQ	3	1	1	—	4	8	Hillingdon House, Uxbridge
Group HQs	3	—	—	—	4	5	By each. Nos 1, 7 and 10
RAF Depot	1	8	4	3	13	26	Based at Uxbridge, Middlesex
Squadrons Inland Area	1	6	4	1	10	16	By each of the eight Squadrons
Central Flying School	1	4	4	2	11	16	Upavon, Wiltshire
Armament and Gunnery School	1	5	4	2	9	15	Eastchurch, Kent
School of Army Cooperation	1	5	5	1	13	14	Old Sarum, near Salisbury
School of Naval Cooperation	1	4	3	1	4	15	Establishment included No 230 Squadron
School of Balloon Training	—	4	2	1	7	9	Rollestone Camp, Salisbury
RAF Station Cranwell	2	11	8	3	22	38	Included Cadet College
RAF Station Halton	3	14	8	4	23	48	Included RAF Boys Training
RAF Base Gosport	1	5	3	2	5	14	Included Nos 3 and 210 Squadrons

The RAF letter plates and typical warning marking on heavy tender bodywork (Peter G. Cooksley).

Unit/Units	Cars, various	Leyland tenders	Crossley tenders	Ambulances	Miscellaneous	Drivers & fitters	*Remarks*
RAF Base Leuchars	1	5	3	2	7	15	Included Nos 203 and 205 Squadrons
Electrical & Wireless School	1	4	4	2	1	13	Flower Down, Winchester
School of Photography	—	1	1	—	1	2	South Farnborough, Hants
School of Technical Training	1	6	3	1	12	17	Manston, Kent
No 11 Wing	1	—	1	—	6	5	RAF Spittlegate (Grantham)
Staff College	1	—	1	—	4	4	Andover, Hants
Flying training schools	1	5	4	2	11	18	Nos 1, 2 and 5 in being
Aeroplane Experimental Establishment	1	5	4	1	9	15	RAF Martlesham Heath
Marine & Armament Experimental Establishment	1	4	6	1	6	18	At Isle of Grain
MT Repair Depot	1	2	1	—	4	13	Harlescott, Shrewsbury
No 1 Stores Depot	1	20	4	1	12	34	Kidbrooke, London SE3
No 3 Stores Depot	1	11	3	1	7	19	Milton, Steventon, Berks
No 4 Stores Depot	1	9	3	1	10	17	Ickenham, Ruislip
Packing Depot	—	20	2	1	9	13	Ascot, Berks
Record Office	1	1	1	—	4	5	Ruislip, Uxbridge
Inland Area Aircraft Depot	2	12	8	2	16	36	Henlow. Also stored MT
Seaplane Training School	1	3	2	1	4	12	Lee-on-Solent, Hants
Instructional vehicles	—	13	12	—	110	—	Personnel as required
Reserve at home	11	39	31	—	77	—	Held mainly at Henlow
Units in Malta	2	4	3	1	6	14	Total RAF on Island
Units, Egypt and Palestine	34	106	100	19	204	537	Plus 39 armoured vehicles
Units in Iraq	40	130	85	62	228	619	Plus 72 armoured vehicles

MT markings regulated

Drivers and fitters were trained at the RAF Depot and in 1925 MT fitters were made responsible for marking their own vehicles. The Air Ministry had decreed in 1920 that the letters 'RAF' would be marked in white on all vehicles held on charge. The letters would be in a central position on both sides of the body of ambulances, lorries, and vans, but for cars only, they would be placed in letters 1½ ins high on the right-hand side of the dashboard.

At the same time it was announced that the chrome yellow painted identification numbers, with their letter suffix, should be obliterated, and all prime movers (except tractors, mowers and rollers, restricted to travel on Air Ministry property), should be registered with the local authority of the area in which the vehicle was normally garaged. Registration had to be completed by January 1 1921 and from that time vehicles were accounted for by their registration numbers and chassis numbers, although only the former would be visible, in accordance with statutory regulations affecting all road users. The following year a change was made to centrally registering vehicles with Middlesex County Council. The RAF made its own regulations as to where the circular licence holding plates would be positioned, as sketches, similar to these officially issued show. During that period most registrations consisted of two letters and a number, the index letters denoting the district licensing authorities of the United Kingdom.

RAF vehicles in India came under different regulations for licensing. As the Indian Government paid for support equipment, the vehicles were not Air Ministry property. Vehicles there displayed number plates front and rear with an Indian Government number, prefixed by a broad arrow to denote Government property.

In March 1922 aluminium plates for the letters 'RAF' in sizes 5 and 3 ins high, for heavy and light tenders respectively, were introduced, with plated screws for affixing them, positioned as diagrams show. Associated with these were stencils issued on the scale of one per three vehicles held on establishment. These were to allow the letters to be polished without adversely affecting the paintwork. It was not only his buttons a driver had to polish in those days of spit and polish.

In overall finish the vehicles of the RAF were the wartime khaki of Army vehicles. In 1930 the khaki canvas of hoods and tilts for tenders was changed to blue/grey and from this time blue/grey became the overall colour of RAF vehicles. RAF cars had black hoods, but these too, from 1930, were changed to blue/grey.

It was not until the RAF expansion was underway that stencils were issued for MT special or warning markings. These were introduced in July 1937 with 2½ ins high characters A to Z and 0-8. That is not a misprint, for the figure 9 reversed made 6. Economy was ever a consideration.

Licensed Leyland, a view from an official sketch promulgated to show the current licence holder position (Peter G. Cooksley).

Nearside markings of a Crossley light tender showing the official RAF positioning for the licence holder (Peter G. Cooksley).

Offside markings of a Crossley light tender showing aluminium RAF letter plates and statutory weight markings (Peter G. Cooksley).

MT loadings and movement

From pre-war days it had been a statutory obligation that vehicles of 15 cwt loading and over should bear an indication of the basic and maximum loaded weight. This had to be marked in not less than 1-inch characters in black or white to contrast. On RAF vehicles this was to be on the bottom offside on the cab. On trailers the weight marking was positioned on the offside front end of the side frame. The Service loading warning of not exceeding 3 tons, marked originally by War Office decree, was continued on RAF heavy tenders until the 30s, as illustrated.

As the standard heavy vehicle, the 3-ton tender was required to carry a variety of stores, including ammunition, but there was a weight limit, under 3-ton capacity, for certain explosives. To give an idea of military loads for which they were needed, the following alternatives are given from the official loading tables: biscuits up to 94 × 74 lb cases; 48 sacks of bread each with 40 × 2 lb loaves; 116 cases of 2-gallon jars of rum; 15,500 sandbags packed in bales of 250; 1,000 blankets stored in bundles of ten; 200 coils of barbed wire; 14 double Marquee tents; 512 gallons of fuel cased in 4 × 2 gallon cans or 700 gallons if in 50-gallon barrels.

The RAF had its own regulations on how many passengers could be carried in its various vehicles, this was spelt out as follows:

Type of RAF vehicle	*Number of passengers*	
	Driving seat	*In coach*
Leyland 3-ton tender	3	30
Leyland 30 cwt tender	3	15
Crossley 30 cwt tender	3	20
Crossley 15 cwt tender	2	12
Morris 20 cwt van	2	18
Morris 25 cwt van	2	16
Morris 30 cwt van	2	18
Ford 1-ton van	2	12
Ford 7 cwt van	2	6
Trojan van	2	4

For unit moves, 50 miles was considered a normal day's MT march, with a convoy speed of 10 mph; 90 miles was considered the maximum distance. An advance party travelling at greater speed would go ahead and make quartering arrangements each night and send one vehicle back to lead in the main party. Spare vehicles were allotted in case of breakdown, when the load would be transferred. A breakdown lorry and crew would bring up the rear. Before the war, it was laid down that there would be 20 yds between vehicles, increasing to 100 yds if it was necessary to put on speed. Vehicles in the UK should travel in blocks of five, leaving good gaps for traffic to pull in when overtaking. The official RAF line on fording depths was motor cycles, 1 ft; motor cars, 1½ ft; lorries, 2 ft; men, 3 ft; horses, 4 ft. The advice on crossing on ice was that 3 ins would bear small parties of men, 6 ins would support light transport and horses, and 9-12 ins heavy transport.

The RAF's vehicles in the early days were certainly not economical to run, as consumption in miles per gallon given below shows:

Type of vehicle	*Capacity (gallons)*		*Consumption (mpg)*	
	Petrol	*Oil*	*Petrol*	*Oil*
P&M motor cycle	1½	⅜	70	1,000
Crossley tourer	18	2½	12	500
Crossley tender	18	2½	14	450
Leyland 3-ton	28	3	6	200

Fortunately fuel was cheap in those days, but all expenditure was most carefully scrutinised. Units, excepting fighter squadrons, and stations kept a log book for each vehicle, *RAF Book 413*, with a record of its maintenance. Fighter squadrons had no authorised transport of their own as it was pooled with their station transport.

Ford Model T and Hucks Starter

Among the most famous cars of all time must be the Ford Model T, produced in America from 1908 and by a factory in Manchester from 1911. The British Services had 18,984 'T' model Fords in use in 1918, a large number of them in

Left *Ford Model T No 913762 with ambulance body on duty at No 37 Training Depot Station, Yatesbury, in 1918. The wording on the side reads: '1 DRIVER AND 4 SITTING CASES OR 1 DRIVER 1 ATTENDANT AND TWO STRETCHER CASES'* (FINA-4581G via J.T.C. Long).

Right *Ford Model T No 2477191 at School of Aerial Fighting, RAF (Canada), Beamsville, Ontario in 1918. Another service Ford T with cab body is in the background* (Public Archives of Canada C30721).

Above *Hucks Starter at Royal Aircraft Establishment, Farnborough, with Bristol Fighter Mark II J6611* (MoD H67).

Left *Hucks Starter at an RAF depot circa 1920* (E.F. Cheesman).

Top right *Hucks Starter and Avro 504K—where aircraft and vehicle make direct contact* (Peter G. Cooksley).

RAF service as passenger cars, ambulances, fire tenders and light tenders. It was in Egypt, Palestine and Mesopotamia that the Ford Ts, known colloquially as 'Tin Lizzies', proved their worth and they were retained for several years in overseas theatres.

There was little difference in principle between the engine of a motor vehicle and that of an in-line engine in an aircraft. In both cases the engine had to be turned, to suck petrol into the cylinders, with the ignition system switched on for the engine to 'fire'. For the motor vehicle there was the starting handle to be turned; a job to be tackled carefully, for many a thumb was broken by a backfire. With an aircraft it was a case of turning the prop, which, when the engine fired, swung round in a manner that cut many a young life short. A careful prop-swinging drill was evolved to avoid such accidents.

An electric starter, such as cars invariably have today, was needed, and firms like Lucas had put their minds to this in the days of the RFC. One idea was a hydrogen starter, and drawings for a trolley for carrying 200 cu ft hydrogen cylinders were approved early in January 1918. The project was dropped in place of electric starters, and these were planned for the Liberty engine of the DH9A and DH10. However, as the apparatus weighed 40 lb, it was finally decided to postpone its installation. This left a need for an external mechanical engine turner and this was met by the Hucks Starter.

B.C. Hucks, a pioneer airman and one of the leading pilots at Hendon 1910-1914, remained at that airfield with the Aircraft Manufacturing Company, which produced aircraft to the designs of Captain (later Sir) Geoffrey de Havilland. When chief test pilot in 1917, Hucks is said to have initiated the idea of a mobile mechanical starter and a senior design draughtsman, a Mr Johnson, adapted such a starter to a Ford T chassis. In 1918, the company, then re-registered as Airco, produced the starter in quantity as the Airco Aero Engine Starter which was marketed by Brown Brothers. Hucks died in the influenza epidemic of the period and the starter was given his name.

In the marketing specification it was stated: 'Aero engines of today (1920) cannot be started by hand without very great risk of accident, the greatest difficulty being experienced in obtaining the requisite compression in the cylinders. The Airco Starter is guaranteed to start *any aero engine* from cold without any doping whatever in thirty seconds, and with no more exertion than is required to crank a standard motorcar engine.'

The starter could be fitted to almost any motorcar chassis, but was generally adapted to that of a Ford T, and the RAF purchased a number of these complete, and some starter mechanisms only. The Royal Aircraft Establishment spent over £1,000 in the experimental fitting of the mechanism to a Crossley chassis.

An Air Ministry instruction of May 27 1920 notified the introduction to the RAF of a 'Starter, aeroplane engine, Hucks, Ford'. It was made mandatory to use these devices for all engine starting, with the exception of Avro 504Ks with 100 hp Monosoupape engines. The operating principle was a rotating shaft, chain-driven from the Ford's engine, with a claw to engage and turn the aircraft propeller. The shaft was telescopic so that the vehicle did not come into contact with the aircraft while it was moving into position.

A starter team comprised a driver and supervising NCO. The NCO would dismount some paces from the aircraft to guide the driver into position and give the command 'All clear' for the driver to extend the claw and engage. The previous routine of hand-swinging drill would then be carried out with the pilot responding to the NCO's 'Switch off (ignition), petrol on'. The starter would give a few turns and then 'Contact' for the ignition switch, and turning would start in earnest, with the telescopic shaft allowing the claw to be thrown back when the engine fired.

The Ford Model T was superseded by later designs in production from 1927, by which time it was used only overseas by the RAF. However the Hucks Starter remained in use until the 30s, when various other methods of starting were in use, including geared winding to turn the engine. Starter motors were introduced, but since the act of starting caused such a drop in battery strength, it was necessary in some cases to use external batteries, and so a new series of vehicles came into use, the starter trolleys.

Chapter 5

Armour for operations between the wars

The decision to use air control as a military force in Iraq, where responsibility had been given to the British Government under League of Nations mandate, led to an RAF Command of eight squadrons and base units being established there. Apart from their domestic transport, at times armoured for travel in areas where local tribes defied the Iraqi Government, armoured fighting vehicles were needed to escort parties who were maintaining emergency landing grounds, across hundreds of miles of desert. They also provided protection for aircraft salvage parties, or for personnel conducting ground searches.

There was also the unhappy burden of Palestine, where the British peace-keeping force under League of Nations mandate was harassed by Arab and Jew alike. Here the RAF formed part of the garrisoning for which an armoured ground element was considered, and proved, to be necessary. Air control was extended to the Aden Protectorate in 1928, but in India the RAF was subordinate to the Army and the armoured ground transport was exclusively Army.

Rolls-Royces and Lancias

The aircraft used in Iraq for air control from 1921 were DH9A bombers, Bristol Fighter army co-operation aircraft, Vickers Vernon (later Victoria) bomber-transports and for the early years a single squadron of Sopwith Snipe fighters—not that any opposition was expected in the air. The ground armour was far more standardised and almost exclusively Rolls-Royce. The RAF took over a number of these wartime armoured cars from the army and placed contracts for more in 1920. The postwar Rolls with its Michelin steel disc wheels, differed visually from the 1914-pattern vehicles, which had spokes.

An armoured car training establishment was set up at Manston in Kent in 1921 under Wing Commander W.H. Primrose DFC. Standard training was in arms proficiency with Lewis and Vickers machine guns, Lee Enfield rifles, Colt automatic pistols, Stokes mortars and Mills grenades. There was also training in Morse signalling to six words per minute, and overhauling and repairing Lancia and Rolls-Royce armoured cars. Some of the rank and file were ex-'Black & Tans'. When the Lancias were first allotted, a party drove them to London for a wild night out, from where the Military Police rounded them up. They were soon shipped out to Iraq under Wing Commander Primrose, arriving in September 1922.

Simultaneously, training had been carried out at Heliopolis from December 19 1921, with ten 1914- and two 1920-pattern Rolls-Royces. This training establish-

Eagle *of No 2 AAC, a typical 1914-pattern Rolls-Royce armoured car chassis, seen in service in Palestine in 1923* (Chaz Bowyer).

ment became No 1 Armoured Car Company (ACC) on February 1 1922. To provide domestic transport for the new unit, a dump of wartime vehicles was raided and six Crossley tenders were made serviceable. With these vehicles the Company moved to Iraq, where drafts from Manston joined them. At that time an RAF company had flight status, but from April 7, when Squadron Leader Alfred J. Currie was appointed to command No 1 AAC, it was raised to squadron status.

During 1922 a consignment of Rolls-Royce armoured cars arrived direct from the factory, supplemented by Lancia armoured cars surplus to Army requirements on the evacuation of Southern Ireland. The Rolls-Royces were issued to No 2 AAC, formed in April 1922 for work in Palestine and Trans-Jordan, with its depot at Sarafand, Palestine. The Lancias, except for a reserve kept at Heliopolis, were sent to reinforce No 1 AAC in Iraq. However, No 1 was disbanded on December 1 1923 to form an Armoured Car Wing at Hinaidi, and the vehicles were re-allotted to form Nos 3, 4, 5 and 6 Companies, except the Lancias, which were switched to No 2 AAC. This left No 2 the largest company, since HQ and A, B and C Sections had three Rolls-Royces each at Jerusalem, Jenin and Amman respectively, whilst the company also had its new section of Lancias, and its repair and transport sections.

For internal security, two spare Rolls-Royce turrets were fitted to two army Dragon tanks by the aircraft depot at Hinaidi in 1924. From 1925, various improvements were made to the Rolls-Royces, including fitting dynamo lighting in place of acetylene lights, and the inclusion of mechanical tyre pumps. In Iraq the armoured car organisation was constantly changing. No 3 AAC was disbanded at Basra on June 1 1925, leaving Nos 4 and 6 AACs at Hinaidi and No 5 AAC at Mosul. Nearly two years later, on April 1 1927, the companies

A section of No 2 AAC Rolls-Royce armoured cars in Palestine on May 9 1923 (Chaz Bowyer).

were reorganised into seven sections of a single Armoured Car Wing, Iraq Command, manned by 24 officers and 200 airmen. The Sections were stationed as follows: No 1 Basra, Nos 2-5 Hinaidi, No 6 Kirkuk and No 8 Mosul.

The cars were frequently used in skirmishes, but in 1927-28 their greatest trial came as part of Akforce, formed when trouble flared on the Iraq-Najd borderland. Two sections of Rolls-Royces operated under Squadron Leader H.A. Whistler DSO DFC, who had been an air ace of the First World War and two further sections came under Squadron Leader S.F. Vincent, who shot down German aircraft in both world wars. At the end of the campaign seven new gearboxes, twenty front road springs and two back axles were issued for the Rolls-Royces, which also needed 125 tubes and 81 tyres (31 × 6) replacing. It was at this stage that the 1914-pattern Rolls-Royces were finally withdrawn. The term 1914-pattern was somewhat nebulous, for the chassis of one RAF Rolls was dated 1910 and others were 1912. These had been private cars, stripped down and re-built, with four tons of armour and extra spring leaves fitted, to a style that in 1914 had set a pattern.

Armed MT

During 1928, a section of Rolls-Royce armoured cars was transferred from Iraq to Aden, where it became D Flight of No 8 Squadron with three flights (A-C) of Fairey IIIFs at Khormaksar. This was on the extension of air control to the Aden Protectorate. Another measure was the raising of Aden Protectorate Levies of locally enlisted personnel under RAF administration and operational control. The initial transport of the original six platoons of Arab troops, largely recruited from the former 1st Yemen Light Infantry, formed in the First World War, was 48 camels and 8 mules.

Above *Typical 1920-pattern Rolls-Royce armoured car, the RAF Type A. A set of spare wheels was carried, and signalling flags were furled on the turret top. Additional four-gallon petrol tanks were stowed externally and baggage and oil and water drums were carried on the running board. On the other side picks and shovels were stowed* (K.M. Robertson Collection).

Below *The 1920-pattern Rolls-Royce armoured cars at the approach of the 40s, when a Lewis gun was mounted for anti-aircraft use* (K.M. Robertson Collection).

The Levies' first standard transport for troop-carrying were Dodge 4 × 4 ¾-ton trucks issued in 1938. They underwent their first serious test in February 1939 when an aircraft force-landed in the Wadi Hassan, 83 miles from Khormaksar. A Levy party set out in six Dodges, the last 28 miles taking ten hours driving along the boulder-strewn wadi. A laudatory report was made on the Dodge trucks, stating that they appeared to stand up to the most exacting service conditions and were proving a most satisfactory vehicle. The day of the Ford was giving way to an era of the Dodge with their vehicles emanating from factories in Britain, Canada and the USA. Many of them were to be used by the RAF at home and overseas.

Ford vans had been retained by the RAF for desert work into the early 30s. By that time the English equivalent of Henry Ford, William Morris (later Lord Nuffield), was making his name manufacturing cheap reliable cars and commercial vehicles. During the 20s, numbers of Morris commercial vehicles were bought by the RAF, which purchased four-wheel general service trucks and ambulances and later the larger 3-ton 6 × 4 trucks. For escort work, some of the Morris trucks were armed with a Vickers machine-gun mounted on the cab roof, on a Scarff gun-ring taken from the DH9A aircraft, which was being replaced by the Westland Wapiti.

Increased armour

Due to the possibility of encountering land mines, in the continuing struggles of Arab and Jew, the Rolls-Royce armoured cars of No 2 AAC in Palestine were fitted with steel armour under the chassis, for the protection of crew and engines, early in 1939. Trials showed the increased weight of 540 lb did not affect the performance of the vehicles to any marked degree. Similar armour was extended to wireless tenders for the protection of the crew and apparatus.

It was evident that the ageing Rolls-Royces would have to be replaced. A Weiss-Straussler was sent for trials at Baghdad in 1935. Alvis acquired a licence for the new car and received an order for 12. The first two cars arrived at No 2 AAC on January 18 1939 for trials. Extensive tests were made on roads, rough ground and the wet marshy areas of Palestine. Selected personnel from the company were given driving instruction while tests continued by operating one Alvis in Trans-Jordan. At the same time two other Alvis cars were being tested in Aden. The Straussler design had first been produced in Hungary in 1933. Alvis substituted a 120 bhp six-cylinder engine of their own design. Since they tended to overturn more easily than the Rolls, the RAF did not increase their original order. Alvis, however, received small export orders from Portugal and the Netherland East Indies.

Another armoured car supplied in limited quantities was the Crossley 6 × 4, but with the Second World War bringing a large increase in army personnel to the Middle East, RAF policing responsibilities were eased and replacement cars were unobtainable. Some Rolls-Royces did operate in the desert in 1940 and the armoured hulls of others were modified for fitting on a Fordson 4 × 2 chassis. The latter were involved in a distinguished action during the Iraqi insurrection. The Fordsons of No 2 AAC were in the Western Desert on airfield protection duties on May 5 1941. Switched to Iraq, they were in action there, 1,000 miles distant, within five days. In February 1942, the Rolls-Royce armoured cars were declared unfit for war and relegated to training. precisely two years later orders were issued for them to be withdrawn and broken up for scrap.

Above *Armoured cars, 1923* (P. Liddle via JMB/GSL Collection).

Below *Weiss Straussler armoured car at Baghdad on trials 1935* (Chaz Bowyer).

Above *Morris armoured tender* Medway *in the early 30s* (Chaz Bowyer).

Below *Alvis-built Straussler at Aden in the late 30s* (Chaz Bowyer).

Above *Crossley armoured car as supplied in small quantities to the RAF in the late 30s* (MoD H2078).

Below *Crossley armoured car fitted for transmitting and receiving wireless messages on the move* (Tom Hayes via P.H.T. Green).

Finish and markings

Armoured cars were finished in light khaki paint as desert camouflage. To assist identification from the air, some of the turrets were painted with large roundels and a small roundel appeared on the armoured sides, near the driver's position, to give due indication that they were RAF, not Army, operated. Rolls-Royce Type A armoured cars were identified by their original War Office numbers W01 to W0280 issued 1920-27.

A particular feature was the naming of armoured cars, painted in capital letters each side of the bonnet. There was no command naming scheme and this was left to unit commanders, usually with a theme and often with capital letters appropriate to their section, A to C. Known names are as follows:

Rolls-Royce *Acasta, Adder, Atalanta, Avenger, Aurora, Bison, Chatham, Cheetah, Cleopatra, Condor, Curlew, Desmond, Eagle, Erberus, Explorer, Golden Eagle, Fox, Hawk, Jaguar, Orion, Panther, Tigris, Shark, Victory.*

Lancia *Beatty, Drake, Fisher, Furious, Polaris, Rigel, Sirius, Watson, Wilson.*

Crossley *Enterprise.*

Alvis *Ace, Queen, King, Jack.*

Armed Morris *Medway, Severn, Thames.*

The names were often prefixed HMAC for His Majesty's Armoured Car. Recognition panels were displayed on bonnets for the benefit of co-operating aircraft dropping messages to the appropriate car when they were dispersed. A formation commander displayed two white balls on a black panel and a section leader a single white ball. For the appropriate flags and pennants, see chapter 21.

Chapter 6

Crossley/Leyland monopoly—Albion/Morris breakthrough

During 1918 the RAF had practically monopolised the complete production of Leyland 3-ton heavy tenders, Crossley 15 cwt light tenders and P&M motor cycles. In the years of stringent economies that followed, the RAF retained these vehicle types as their standard transport and the three firms concerned received orders for spares, and in some cases overhauls, but with the large wartime stocks which the RAF held, there were no orders for replacements. When, in the late 20s, the RAF eventually required replacements, they shopped around, keeping the new ranges offered by Crossley much in mind, but also looking hard at what the market could offer.

The Crossleys

The brothers F.W. and W.J. Crossley had been motor manufacturers for some ten years when the RFC took their transport to France, as part of the British Expeditionary Force in August 1914. As the Crossley light tenders and touring cars then used by the RFC proved equal to active service conditions and the rough byways and paving in Northern France and Belgium, standardisation was sought on Crossleys, by replacing other types in use and equipping newly mobilising squadrons with Crossleys. As a result, practically the whole of the output of motor vehicles of this Gorton, Manchester firm during 1914-18 went to the RFC, or RAF, as the latter had become in April 1918.

With some 80 squadrons behind the 90 miles of Western Front held by British and Commonwealth forces, plus kite balloon sections and depots and an expanding Independent Force all using transport, the RAF built a large motor repair centre at Motteville, on the railway between Le Havre and Rouen, in which the Crossley section alone comprised a 337 × 75 ft structure with outbuildings for blacksmiths, tinsmiths, welders and a Crossley office block.

Barely had the new depot got into full swing, than the Armistice came and with it the rundown of the RAF. Most of the Crossley tenders and tourers used in France were brought back for disposal. The Motteville depot was kept busy well into 1919, reducing worn and damaged vehicles to scrap metal, but passing 534 tenders and 101 tourers for return to the UK, where a surplus already existed from the contraction of training units.

For the peacetime Force, a total of 250 tenders was required for use at home stations. These were selected from stocks being held at Henlow and those currently in use and were sent to Gorton for complete overhaul in 1920-21. In fact this was a complete breakdown and rebuilding from spares and parts of other vehicles. As a result, the firm cancelled the original chassis numbers and issued new numbers in a 611-1809 range known as their 'PN' number, which became their official RAF identification number. The five main components—motor, gearbox, rear axle, front axle and steering—were each recorded by their individual number, except where they were standard War Office spares. The precise number of Crossley 20/25 hp engines Nos 2446-12305 (a total of 9,860) was supplied to War Office orders, as were 11,387 gear boxes, an item on which there was undoubtedly heavier wear.

The Crossleys, tenders and tourers, which used the same engine, were finished in khaki until later in the 20s, when the RAF adopted its blue-grey, but hoods were supplied from the outset in grey canvas. An exception were six RAF Crossley landaulettes used by senior officers at the Air Ministry and one allotted to the French Aviation Mission, which remained in highly polished commercial finishes.

During wartime years a feature of the Crossleys had been their double wheels on the rear axles. Under peacetime economies these were reduced to single wheels, except for those on active service in India and Iraq, and in Ireland up to 1922, when the RAF withdrew from Southern Ireland on the inauguration of the Free State.

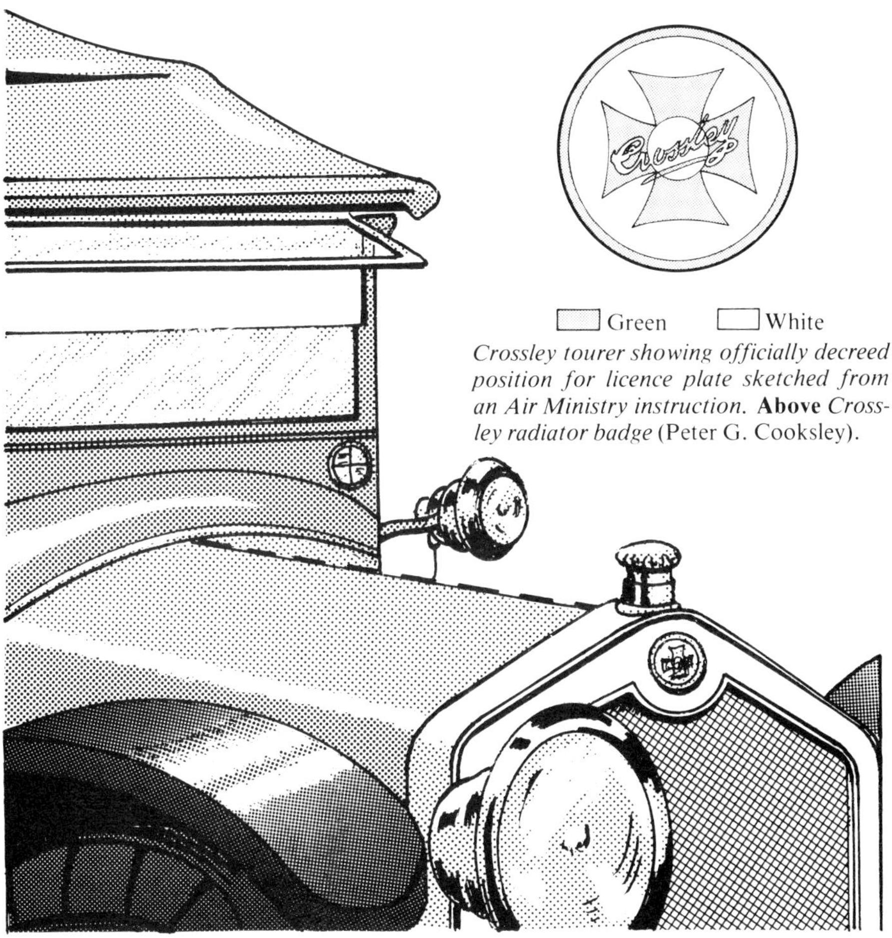

Crossley tourer showing officially decreed position for licence plate sketched from an Air Ministry instruction. **Above** *Crossley radiator badge* (Peter G. Cooksley).

Above *Crossley tender on the road to Baghdad, circa 1918-19* (LPC D1682).

Below *Morris Commercial 6 × 4 tender being refuelled from an RAF station pump. The notice on the pump is not a brand name, but an 'MT—DANGER—PETROL' warning* (MoD H499).

Ten years after the war, the RAF had to review the possibilities in replacing their ageing wartime transport. A limited number of 19.5 hp Crossley staff cars were bought to replace tourers. New Crossley 30 cwt fire tenders were bought and some special Crossley-Kegresse half-tracks for use in desert regions. Acquisitions of Crossley types in the years between the wars for service or evaluation were as follows:

Type	*Description*	*No*	*Chassis numbers*
A	25/30hp Type IX chassis	92	15211-15302
B	19.5hp 4-cylinder car	115	17051-17165
C	18/50hp 6-cylinder car	1	40046
D	20.9hp 6-cylinder staff car	48	41397-41444
E	25/40hp 1GL1	48	30072-30118
E1	25/40hp 1GL1 half-track (Kegresse chassis)	4	30061, 30064 30067, 30071
F	25/40hp 1GL2 four-wheel	65	30301-30325 30501-30540
G	25/40hp 1GL2 six-wheel	2	1146/3 and 1146/4
H-L	30/70hp 1GL3 six-wheel with wireless or tender body wooden or metal	100 (approx)	30701, 30808-38, 30901, 31121, 31334-83, 31501 +
P	Fire tender body	50 (approx)	Nos not traced

Crossley tourer at the Front with Captain J.R. Hopkins DFC of 'B' Flight, No 21 Squadron.

The Crossley Company ceased car manufacture in 1937 to concentrate on commercial vehicles and while numbers of these were to be purchased in quantity during the Second World War, they were no longer exclusive suppliers to the RAF. The firm was absorbed by a merger of AEC, Maudsley and Thornycroft as ACV (Associated Commercial Vehicles) which in 1948 merged with Leyland.

The Leyland monopoly

During the First World War, 5,932 Leyland lorries were built, 4,721 of them to a standard known as 'RAF Type A'. At its wartime peak in late 1918, the RAF had 6,500 Leyland tenders in use, as almost the whole Leyland lorry production during the war had been for the RFC, and then the RAF. Added to this were Leylands taken over in 1914 for the Army under the War Office subsidy scheme, and others impressed, and allotted to the RFC. As with Crossleys, so with Leylands, a large covered area, with outhouses and an office block, was built exclusively for their repair in the Motteville RAF vehicle repair depot.

As well as being standard squadron equipment, RAF Leylands were the standard tenders for depots and the three Air Armament Columns serving the Independent Force. Tenders were of two main types; standard trucks with collapsible side seats, and floats with a floor extension to take long objects such as wings and fuselages. When, in 1930, a car ran into the side of a rear extension of a Leyland at a cross-roads, orders were given for the extension sides to be painted white, 20 ins from each end. A Leyland's cost was rated at £510 by the Treasury, about half the cost of a aero engine of the day. Apart from carrying personnel and stores, and salvaging aircraft, Leylands were used as workshops, mobile telephone exchanges, photographic laboratories, mobile chart rooms and balloon-winch-carriers, and therefore had various types of bodywork.

Postwar Leylands served in Aden, Egypt, India, Iraq, Palestine, and the UK. Up to mid-1922, when the Irish Free State was formed, the whole of Ireland was within the United Kingdom, with Nos 19, 22, 23, 24 and 25 Training Depot Stations at Curragh Camp, Gormanstown, Baldonnel, Collinstown and Tallaght respectively. Apart from these, there were coastal stations and aerodromes for squadrons mobilising for France, with each establishment having a quota of Leyland tenders. In the Irish troubles of the early 20s, pending the establishment of the Irish Free State, some Leylands were partially armoured and travelled in convoy on duty runs between ports and stations. It says much for the tension of the times that in Dublin, when a Leyland tender backfired, it was attacked and set on fire because the bangs were thought to be shots!

Sponsored by Leyland in 1922, Trojan Ltd started building light vans at Kingston-on-Thames in the former Sopwith works which Leylands had acquired for reconditioning ex-RAF tenders. A small batch of Trojans was the first group of light utility vehicles in RAF service. Powered by a 1,488cc duplex two-stroke engine under the seats, with chain drive to a rear axle without differential, they were economical both to buy and to run, which endeared them to the Treasury of the day. The distinctive pop-popping, of their engines, heralding their coming, was one of the sounds of the late 20s, when the majority of those around were either in RAF blue, plying between railway and RAF stations, or in the red of their prime commercial operator, Brooke Bond tea. The firm moved to Croydon later and Hawker Siddeley took over their former works.

Crossley tender registered AM6701, marked on the cab side CFS No 5 for Central Flying School No 5 vehicle (P.H.T. Green).

In the early 30s the RAF tried to increase the life of its Leylands by arranging for Dunlop to replace the solid tyres of the Type A with wheels taking pneumatic tyres. These Leylands were officially known as Dunlop pneumatic conversions. As late as 1934 these were being fitted with pumps on the side of the cab, but their days were numbered. The RAF decided against the modern Leyland vehicles which were finding favour with the Army and went to Crossley, Morris and Albion for their standard and specialist transport. Leyland, as we now well know, survived and went on to expand in a series of postwar mergers.

Morris, Albion and Bedford

The breaking of the Leyland/Crossley monopoly started in the mid-20s with William Morris, who, having had success in introducing car mass production to Britain, commenced the output of commercial vehicles under the trade name Morris Commercial. By 1927 the oldest of the Crossleys and Leylands were being withdrawn and replacements were needed. There were two reasons for orders being placed for Morris 6 × 4 trucks at a time when limited numbers of new Crossley 6 × 4 trucks were on order. Firstly, the Morris had the appealing aspect of cheapness and secondly it was becoming policy, as war again became a possibility, to avoid relying on a single source of supply. Both Morris and Crossley types had various bodies fitted for refuelling, ambulance and fire tenders. In all cases the Crossleys proved far superior in performance.

When Albion brought out their AM463 4 × 2 truck in 1934, it came as a suitable replacement for the Crossley light tenders with truck, ambulance and other bodywork. It was greatly welcomed by drivers, for it had the then refinement of an enclosed steel cab with cushioned seats and it incorporated a self-starter. In general it was quieter than the Crossley, except that the steel cab was liable to vibrate. With the expansion, orders were increased and the Air Ministry took the majority of the 1,900 built.

Another factor which had a great influence on RAF MT, and still has, was the rise of Vauxhall. Manufacturers of cars since 1903, the company was taken over by General Motors of America in 1925 and using the trade name Bedford, went into the commercial vehicle market, which they dominated by 1939.

Above *The Albion AM463 with enclosed metal cab, in this case with 450 gallon fuel tank bodywork* (K.M. Robertson Collection).

Below *Leyland heavy tender with War Office number in RAF service, shows its odd wheels of three different types* (M. Aldridge).

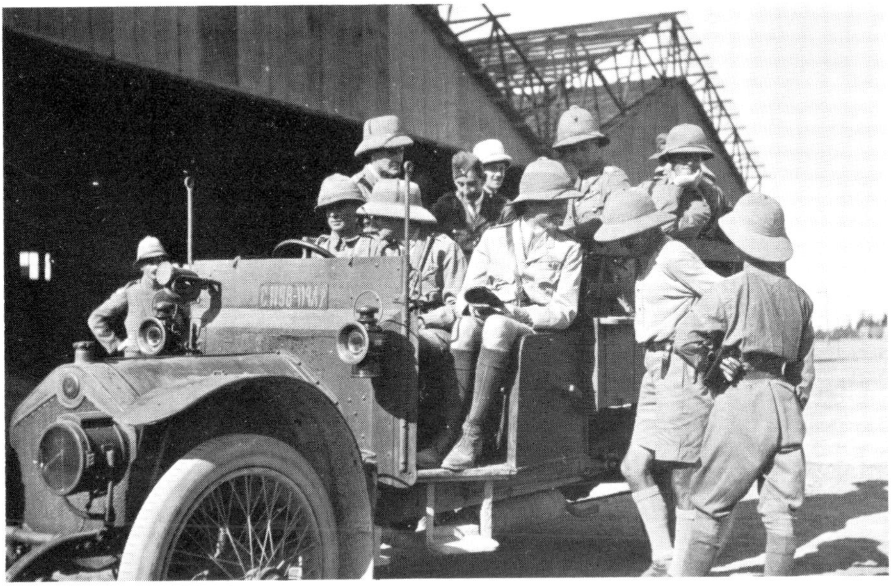

Crossley tender C1198 (number also repeated in Arabic) of No 14 Squadron in Mesopotamia in 1918 (E.F. Cheesman).

Expansion and war

The RAF expansion from 1934, with new squadrons forming and new stations opening, meant a large increase in the numbers of vehicles required. No 1 MT Storage Unit was opened at Hendon and moved in 1936 to Wembley, where the RNAS had once experimented with armoured fighting vehicles; later No 2 was opened at Whyteleafe and No 3 at Norwich. With new MT deliveries being in advance of aircraft production, two MT store sub-depots were opened at Park Royal and Pulham. This station had been the former Kite Balloon Experimental Station. Further stores depots for MT were opened, but in 1936 their names were changed to Equipment Depots and the following year to Equipment Units and again to Maintenance Units.

When war came, orders were increased and many vehicle types came into use by the impressment of civilian vehicles, although the army had first call on these. At one period, early in the war, while the five maintenance units for MT held 750 vehicles each, storage space was still required for 9,050 vehicles, resulting in many being temporarily stored in the open. By the end of the war, in 1945, the stored reserve totalled 26,320 vehicles.

Chapter 7

Wartime vehicles: their colouring, markings and type numbers

Strange as it may seem, wartime was the most colourful period for RAF MT as regards variety of colouring and changes, albeit in rather drab shades. Regulations had conditioned the markings in the years of peace to an overall blue/grey and highly polished surfaces. The war brought radical changes.

No longer could the RAF call upon the proven suppliers of their formative years to meet the greatly increased demand. Leyland, apart from a heavy commitment for Army vehicles, went over to tank production, turning out Covenanters at first, followed by Churchills, while their foundry met RAF requirements in other ways by producing 250 lb HE bomb casings and later 1,000 lb bomb casings. Morris Motors not only went over to Crusader tank production, but the large-scale production of Tiger Moth training aircraft.

Squadrons moved by road and only their aircraft went by air, except in emergency or special practice moves, when personnel and equipment flew in Bombay or Harrow transports. On other occasions there were vehicle/aircraft interchanges. For example, early in the war, in November 1939, it was decided to standardise on three types of airborne radio transmitters/receivers. Plans were for aircraft to fly into St Athan for re-fitting as necessary, but at times that station's airfield was unserviceable and fitting parties were quickly formed to

travel by MT round airfields. Later in the war, with electronic sophistication, many radio/radar fitting parties and special servicing teams were mobile either by air or road.

To rationalise transport demands, a number of RAF MT transport companies were raised in 1941. These vehicles were distinguished by their company number displayed on the wing. In the immediate postwar years these companies were reduced to four, operated by No 54 Wing under the direct operational control of the Air Ministry, except No 2 Company at Leicester East, Stretton Road, Oadby which was operated by Maintenance Command. The other three companies were No 4 Company at St John's Wood, London; No 5 Company at Woodvale, Formby, Liverpool and No 6 Company at Bishopbriggs, Glasgow. The RAF's main MT Allotment and Storage Section was located at Bicester, Oxfordshire.

Specialist vehicles proliferated and so did aircraft-carrying vehicles. The four-engined bombers brought a need for long transports, and even so, it took six such vehicles to convey one bomber. By the end of the war the MT Salvage Fleet alone consisted of 750 low-carrying trailers, 170 cranes, 357 3-ton tenders, 47 15 cwt trucks, 78 light vans and 96 cars.

Camouflaging MT

The outbreak of war did not have any immediate effect on the finishes of MT, for their standard blue/grey uniform finish did not render the vehicles particularly conspicuous. It was different with aircraft. Their silvery aluminium doping was replaced with a disruptive-patterned dark earth and dark green camouflage from 1937 onwards.

Some changes for MT came with the Battle of Britain, which brought daylight attacks on airfields. It was argued that, while the aircraft in their green/brown appeared to merge with the landscape, the uniform colouring of vehicles near

In the early war years when RAF vehicles bore the blue colour similar to RAF uniforms. At this Whitley bombing-up scene are an Albion tender and Humber Snipe heavy utility (K.M. Robertson Collection).

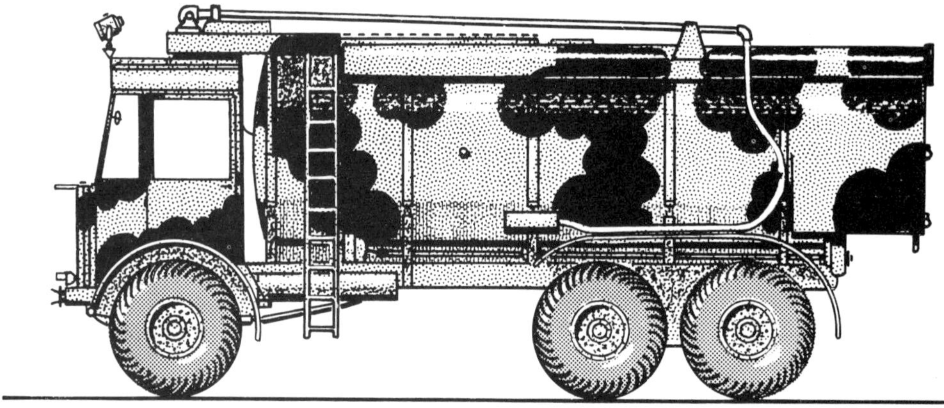

Above *'Mickey Mouse ear camouflage pattern' as applied to AEC 2,500 gallon refueller* (Peter G. Cooksley).

Below *The 1944 olive drab finish has been applied over the black and olive green disruptive pattern camouflage of these vehicles of a No 84 Group radar unit in convoy in France 1944-45. The large roundel on the side of the vehicles was a marking peculiar to MT of the 2nd Tactical Air Force* (E.F. Cheesman).

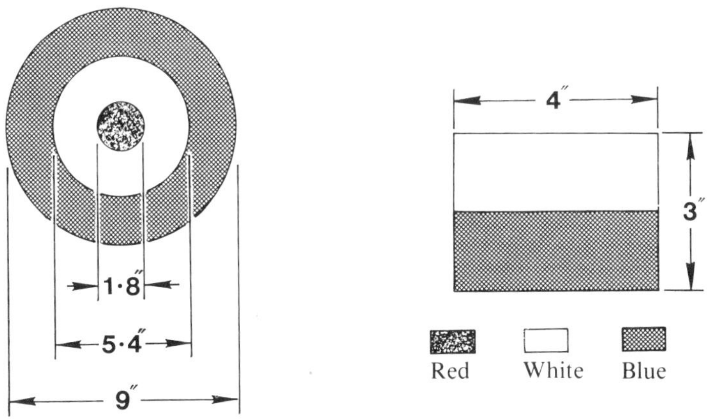

Left *RAF standard MT roundel sizing.* **Right** *Air Formation Signals flash* (Peter G. Cooksley).

them compromised their camouflage. For stations in South-East England it was decided that a disruptive-patterned camouflage similar to that of aircraft should be applied, but the dark earth and dark green paints, applied on production, were only held in limited quantities by stations for repair work. However, large quantities of industrial camouflage paint were held for the concealment of hangars and aerodrome buildings and these colours were applied.

It was not until August 1941 that a standard camouflage scheme for RAF vehicles was introduced. It was applicable to all vehicles on charge, including those impressed, but excluded those hired, or cars issued to officers of air rank (ie Air Commodore and above). Since the weight problem was not so acute with MT as with aircraft, the vehicles were painted overall in khaki green No 3, with certain areas overpainted in Nobel's tarmac green No 4 to give a disruptive pattern.

In adopting this new camouflage scheme the RAF was influenced by the Army. Between the wars, Army vehicles had been a single colour—a semi-matt dark green—until February 1939, when khaki-green No 3, the Army's standard colour for webbing blanco, was notified as the shade for all vehicles, excepting limousine staff cars, but including field guns and equipment. It was coincidental that on May 15 1940, five days after the German offensive opened in the west, leading to the Dunkirk evacuation, the War Office adopted a disruptive pattern of dark tarmac over khaki green for vehicles and artillery pieces. Since vehicle production from that time was being controlled for all Services by the Ministry of Supply, a standard finish was of some importance, and from that time until after the war, the finish of RAF vehicles was similar to those of the Army.

The green and tarmac colours were a compromise including a field colour for open country and a tarmac road colour. There was no connection with the so-called tarmac apron in front of hangars, as these were often of concrete and of a much lighter colour, and some aprons were sprayed with industrial paints to tone them down. However, most perimeter tracks were actual tarmac (tar and Macadam), so the Army colour was also ideal for RAF vehicles.

Anti-gas precautions were much in evidence, particularly in the early war years. A patch of gas detector paint, not less than 6 × 6 ins, had to be visible to drivers from their cabs. This was not practicable with flat-fronted vehicles, but

No less than 13,000 Austin K6 3-ton trucks were built during the war. This example, RAF99798, adapted to carry a Type 7 radar array, bears the coding H27 on its wing, the letter for its command (Technical Training) and the number for No 27 Group (K.M. Robertson Collection).

for the majority, the top of the offside mudguard was the usual position for this patch. An exception was American vehicle types with left-hand drive, then arriving from ex-French orders and under Lend/Lease. The patch on these would be placed on the opposite wing. The anti-gas patch was a lemon yellow colour until 1941, when brown anti-gas paint was introduced. While originally displayed as a square, it was later a blotch befitting the camouflage pattern.

After September 1943, the need for gas detector patches diminished, as such paints were introduced into the production of overall finishes. Simultaneously, there was a change in shades to paints specified as MT Brown Special and Matt Black. The black was introduced because vehicles were often placed near buildings, at the side, in shadow, to aid concealment, and in such cases the black would tone in, and the brown, if conspicuous, would not have the rectangular form of a vehicle top. The black was patterned in large lobes giving rise to the finish being facetiously called 'Mickey Mouse Ear camouflage'. Paints were usually sanded to give a matt finish, but after weathering the black soon became a mid-grey. For some inspections this was rubbed over with an oily rag to give it a slight sheen and smarten up its appearance.

For the invasion of the continent, planned for late spring and effected in early summer, this finish was considered too much of a winter camouflage, and olive drab replaced the brown. This was the position for D-day, when many American vehicles supplied under Lend/Lease were in overall olive drab. This may have influenced the Army and Air Force to standardise on olive drab, which, as a shade, was first introduced as a British Standard War Emergency Camouflage Colour in 1944. Judging from the BSI colour sample, it was more drab than olive. The Army issued their instructions in August 1944 and the RAF the following month. Vehicles were to be finished in the new colour only when repainting was necessary; there was to be no painting merely for the sake of uniformity, so it was the spring of 1945 before olive drab was the general colour for most RAF vehicles in the North-West European Theatre.

It might have seemed helpful to modellers to have provided a table of changes with dates, but this could be misleading because of the time lag between instructions being promulgated and carried out. Not only were there the provisos on delaying changes until repainting became necessary, it was often a case of using up stocks of paints before effecting changes. Uniformity was, in any case, the antithesis of camouflage.

Identification markings

Licensing of Service vehicles was suspended in January 1941 and civil number-plates were then generally removed, with the exception of those integral with rear lights. RAF vehicles were identified by a service-registered number prefixed RAF. The RAF roundel was adopted as a Service marking in April 1941, to be positioned on the offside of the front of RAF vehicles, where possible on the wing. Roundels were hand-painted using stencils until 1944, when a roundel transfer was introduced. Also in April 1941, a Command and Unit identification letter and number were introduced. In general, the letter was indicative of the command, and the number was the group number or a code number for the unit. This was marked on the front of vehicles, on the opposite side to the roundel, in 5 ins high white characters.

Letters allotted, identifying commands and formations were as follows:

A Army Co-operation Command (to June 1 1943)
A Airfield Construction Service (from 1943)
B Bomber Command
C Coastal Command
D Allied Expeditionary Air Force (from 1944)
D Middle East Air Forces (to 1944)
F Fighter Command (including the period November 43–October 44 when it was known as Air Defence Great Britain)
H Technical Training Command
K Balloon Command.
L Air Transport Auxiliary, Air Training Corps and University Air Squadrons
P Petroleum Board
Q HQ Transport Command Planning Staff (1944 only)
R No 60 Group
S Colonel Turner's Department (cover name for camouflage and decoy services)
T Flying Training Command
V No 44 (Transport) Group
W Air Ministry Works Department
X Colonel Turner's Department (additional code letter)
Y 2nd Tactical Air Force (1943 only)
Z Ministry of Aircraft Production

In MT transport companies the company number was displayed on a wing of their vehicles. A refinement in No 8 Company was to have the circles of the figure eight colour-coded for each section of the company.

Specialist markings

Air Formation Signals, working closely with the RAF in the field, belonged to the Royal Corps of Signals, and RAF vehicles attached to AFS units bore the Signals flash on their vehicles as illustrated.

Bomb disposal in general throughout the Services was the responsibility of the Royal Engineers, with the co-operation of the Royal Army Ordnance Corps. But bombs falling from RAF aircraft by accident, or from enemy aircraft on Air Ministry land, were the responsibility of the RAF, who were also responsible for salvaging their own wrecked aircraft, which sometimes entailed a task of bomb disposal. From the mid-40s, when bombing started in earnest, the RAF bomb disposal organisation was expanded. Bomb disposal squads, each with an area of responsibility, consisted of an officer, Flight Sergeant, two Sergeants, junior NCOs and men, totalling about 25 per squad. Their transport was standard 3-ton and 15 cwt trucks, utility vehicles and motor cycles. The marking of their vehicles was non-standard. They were painted with bright red mudguards and bore a yellow plate with the words BDS for Bomb Disposal Service.

RAF police, while using standard vehicles, were permitted to carry an 18 × 6 ins board with the words 'RAF POLICE' on it.

Bridge classification markings

In the field, the RAF had to rely on the Royal Engineers to make routes road-worthy and the Pioneer Corps (who gained their Royal title for their work in the war) to maintain them. The Engineers built their bridges in ton weight classifications in the range: 5, 9, 12, 18, 24, 30, 40, 50, 60, 70, with the appropriate number being indicated at the bridge approach. Vehicles were similarly marked with their maximum loaded weight, so that it would be known that it would not be safe for vehicles to cross the bridge unless the approach sign bore a higher number than that on the vehicles.

At home stations, the RAF had little need to be conscious of bridge markings, but in the field, using roads that the Army had cleared, their vehicles had to

MT bridge classification sign styles (Peter G. Cooksley).

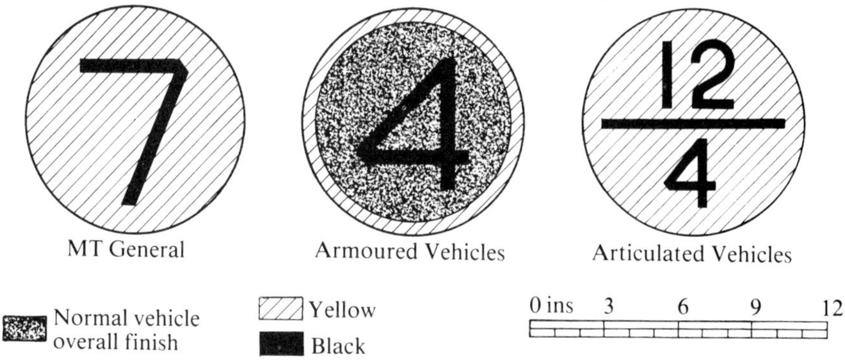

correspond to Army regulations on this point—and it is still a matter of Joint Service agreement. Early in 1944, preparatory to the invasion of the Continent, the Air Ministry regulated the bridge classification markings for RAF vehicles. They were to bear a 9 ins diameter disc, painted yellow, with the maximum tonnage in black figures. In the case of trailers this was to be marked at the front. With articulated and towed vehicles two figures were given, one above the other, as illustrated; the top figure being the total weight of the whole vehicle, the lower figure the weight of the tractor unit alone, or with normal towing the figures would relate to prime mover and trailer. To avoid compromising the camouflage of armoured vehicles, which might be used in a reconnaissance role, there was just a yellow ring of 9 ins diameter to contain the black number.

Examples of MT bridge classification numbers allotted to RAF vehicles in 1944 are as follows:

Class	*Type of vehicle*
2	Hillman Minx saloon
3	Humber Super Snipe saloon
4	Morris Mark I reconnaissance car
5	Albion AM463 heavy ambulance
6	Bedford 3-ton 350 gallon tanker
7	Eagle 5-ton heavy bomb trailer
8	Leyland Retriever 6 × 4 3-ton truck
9	Crossley 4 × 4 workshop lorry
10	Radio vehicle Type 460
11	Ransomes & Rapier excavator
12	Dennis Max 6-ton truck
13	Thornycroft Amazon with 3-ton crane
14	Michigan TMCT-16 excavator
15	AEC mobile oxygen plant
20	Caterpillar D8 tractor
35	Mack 6 × 4 18-ton transporter

Type number marking

With Britain poised to invade the Continent, planning staffs were concerned with the logistic support of the venture. To facilitate the ordering of the correct spares for vehicles, when such a diversity of types with varying designations existed, it was decided early in 1944 to type number the various types of vehicles. Numbers were allotted in blocks according to the class of vehicle, with 1-999 reserved for specialist signals vehicles, 1000-4999 for general prime movers and from 5000 upwards for trailers. The type number was to be painted in white 3 ins characters in the style 'TYPE 1000', or with the number placed beneath 'TYPE'. This was to appear on the nearside (ie left) cab door of prime movers and in a conspicuous position on trailers.

Representative type numbers are tabled with the types to which they refer, as an indication of the great diversity of vehicles operated by the RAF.

No	*Vehicle*
100	Twin channel VHF R/T transmitter
105	Mobile D/F (direction-finder)
150	Twin channel VHF receiving station
1000	Articulated tractor
1100	6-ton transporter
1101	Extra heavy crane
1102-3	Cranes, heavy and light
1104	Oxygen plant tender
1105	Parachute dryer tender
1106	Parachute servicing tender
1107	2500 gallons fuel tanker
1108	Snow clearance tender
1109-11	10-, 15- and 20-ton tenders
1200	5-ton tender
1201	Balloon-winch tender
1202	Bomb-carrying tender
1203	Hydrogen cylinder tender

No	*Vehicle*
1204	Pantechnicon
1205	Sullage tender
1206	Tipping tender
1207	Air Compressor tender
1208	18-ton transporter
1300	3-ton standard tender
1301	Fire/crash tender
1302	Flying control caravan
1303	Mobile floodlight
1304	MT breakdown tender
1305	Mobile office
1306	Mobile operations room
1307	1,000 gallons fuel tanker
1308	2,000 gallons fuel tanker
1309	Photographic laboratory
1310	7 kW power vehicle
1311	Power vehicle
1312	350 gallon water tanker
1313	Workshop lorry
1314	Recruiting van
1315	Sound channel vehicle
1317	Photographic power and water tender
1318	670 gallons water tanker
1319	Mobile dental surgery
1320	Mobile dental workshop
1321	2,500 gallons petrol tanker
1322	Bomb carrier
1323	350 gallons light fuel tanker
1324	Mobile hydrogen compressor
1325	300 gallons water tanker
1326	Catering tender
1327	Stores vehicle
1328	500 gallons water tanker
1329	550 gallons fuel tanker
1330	3-ton tipper
1400	30 cwt light tender
1401	Pre-heater van
1500	15 cwt van
1501-2	Bread/meat stores vans
1503	Welding equipment van
1504	Navigation trainer van
1505	Bomb-sight servicing van
1600	5/10 cwt van
1601	Letter sorting van
1603	Air Officer Commanding car
1604-5	Heavy/light passenger cars
1700/1	Heavy/light ambulances
1800	Armoured cars

No	*Vehicle*
1900	5 cwt 4 × 4 car
2000	Utility heavy passenger car
2001	Touring car
2002	Utility light passenger car
2100	Passenger coach
2200	Mobile store freighters
2201	Domestic fire tender
2300	Mechanical horse
2400	Petrol and oil refuellers
2500	200 gallons water tanker
2600	Light tractor
2601	Light tractor with winch
2700	Heavy tractor
2701	Heavy crawler tractor
2702	Works tractor
2703	Semi-tracked tractor
2800	Platform elevating truck
2801	Works truck
2802	Stacking truck
2900	Motor cycles
2901	Motor cycle combinations
3000	Ditchers
3001-2	Dumpers Marks I and II
3007	Hi-lift shovel
3008	Paver concreter
3009	Power graders
3010	2½-ton roller
3011-2	5/6 and 8/10-ton rollers
3013-6	Tractors classes I-IV
5000-2	Articulated trailer types
5003	Air bottle trailer
5004	Air compressor trailer
5005	Photographic servicing trailer
5006	Film processing trailer
5007	Photo multi-printing trailer
5008	Film office and store trailer
5009	Torpedo-carrying trailer
5010	Workshop trailer
5011	7-ton articulated trailer
5012	2,000 gallons fuel tanker
5100	Flat two-wheel trailer
5101-2	5- and 8-berth caravans
5103	Caravan with astrodome
5104	Dental caravan
5106	6.3 kVA power unit trailer
5107	MT servicing trailer
5108	150 gallon water tanker trailer

No	Vehicle
5109	1-ton tipping trailer
5110	Air compressor trailer
5111	High pressure Jenny trailer
5200	Two-wheel trailer
5201	Towed aerial lighthouse
5202	Balloon-winch trailer
5203	Bomb-carrying trailer
5204	Cooking trailer
5205	Mobile decompression chamber
5206	Dental trailer
5207	Floodlight trailer
5208	Hydrogen cylinder trailer
5209	Landmark beacon trailer
5210	Mobile office trailer
5211	Oxygen plant trailer
5212	450 gallons oil tank trailer
5213-4	450/900 gallons petrol trailers
5215	Photographic processing trailer
5216	7 kW power trailer
5217	Mobile store trailer
5218	300 gallons water trailer
5219	Woodworking trailer
5220	Workshop trailer
5221	500 gallons oil trailer
5222-3	14kW/20kVA power trailers
5300	Flat eight-wheel trailer
5400	Towed excavator transporter
5401	Towed fuselage transporter
5402	Mechanical horse trailer fittings
5403	Mobile parachute drier trailer
5404	Mobile parachute packing trailer
5405-6	Types A2/B2 silicon trailers
5407	Towed tipper
5408	Mobile works laboratory trailer
5409	Works truck trailer
5410	25/25-ton works trailer
5411	Workshop plant trailer
5412	3-ton recovery trailer
5413	Mobile 60 kW generator
5500-2	Air compressor trailers
5504	Bit tar mixer trailer
5505	Mobile boring rig trailer
5506-8	Concrete mixers
5509	M20 type towed crane
5510	AD3 type towed crane
5511	5-ton crushing plant trailer
5512	10-ton crushing plant trailer
5513	Drying unit trailer
5514	Mobile 18 kW generator trailer
5515	Portable belt conveyor
5516	Portable screens and bins
5517-8	K-30/H3 rooters
5519-22	Scraper types
5523	Tar boilers
5524	Water sprayers
5525	Wheeled blade graders

Type indication was yet another marking on these much embellished signals vehicles. The nearest vehicle, Crossley 3-ton RAF134121, bears the unit number 15071 and the chalked '5' is its place in convoy. Also on the front is 'TAF84' for Tactical Air Force, No 84 Group. The next two vehicles are Austin WOT1s with Type 150 and 100 bodies for VHF receiving and transmitting (E.F. Cheesman).

The painting of type numbers was notified as being no longer necessary on September 14 1944 as it was insufficiently precise to ensure the correct spares for all variants. While the painting of these numbers was discontinued from this time, there was no general instruction to delete the numbers and they remained on many vehicles until the end of the war. An exception to the above was made in the case of Radio Vehicle Types (RVTs), for from December 7 1944 it was made mandatory for these vehicles to bear in white 3 ins characters on the left-hand cab door, and on the forward side of trailers, the word 'TYPE' followed by the type number in the 1-999 series.

A picture that tells it all, even showing the placing of the RAF identification number at front and rear, the roundel placing and the dual bridge number for the trailer it tows seen on the left behind the Halifax bomber. The 'B' under the RAF number is the Bomber Command formation letter, but in this case the Group number has not been marked (IWM CH4464).

Chapter 8

Theatres and campaigns

In the previous chapter the wartime markings of RAF vehicles at home and the general standard for factory finishes were given, but in the various theatres of war other regulations appertained.

There was another difference overseas; the drivers were practically all male. Only at home was there a large-scale substitution of WAAF (Women's Auxiliary Air Force) drivers and then at first, with complete disregard for the equality of the sexes, the substitution rate was five WAAF drivers to replace four RAF, but in the spring of 1944 it was regulated to a one for one basis.

France 1939-40

To support the British Expeditionary Force in the field in France in 1939 the RAF provided an Air Component of 13 squadrons and an Advanced Air Striking Force (AASF) of ten squadrons, but while the squadrons of the Air Component were mobile, their ancillary units were not, and in the AASF neither the squadrons nor supporting units were mobile. Such standard RAF vehicles as were shipped from the UK were for domestic support. So great was the shortage of utility vans for normal liaison and communications, that 40 Renault vans were bought locally in France, plus a few buses for central holding. A particular shortage was a lack of 3- to 5-ton cranes.

Early in 1940, a scheme was implemented to increase mobility, but when the German onslaught came in May 1940, the force was still 750 vehicles under strength, the most serious deficiency being 300 3-ton vehicles below establishment. These deficiencies became even more acute with the reinforcements of personnel and equipment by air, since they depended for ground movement on the existing over-strained MT.

The vehicles, subject to home station regulations up to September 1940, were in many cases painted a disruptive pattern of khaki green in the field. On most of the vehicles the British civilian number plates had been removed and identification was by RAF number. Security was very strict and no unit markings, not even small roundels, were permitted. Unit identification on vehicles was by a coded system of numbers for administrative use and postal convenience. This consisted of three-digit numbers, prefixed 'A' to denote the Air Force series, displayed in 2 to 3 ins characters on cab sides. Weights for the larger vehicles appeared in convenient positions, low on the bodywork, for shipping details.

When the evacuation of the BEF began, there were 7,000 troops attached to the RAF who were only 50 per cent mobile. The majority were evacuated to port areas by train, under RAF arrangements, and for the rest the RAF had to loan

Above *A well-concealed Photo Section of a squadron in France, 1940, with Bedford WTH photo laboratory trucks and a French interpreter's car* (Cosford A0598/18).

Below *RAF Fordson 3-ton 6 × 4 truck and heavy utility in middle stone desert camouflage. Gladiators in formation overhead* (Newsreel).

100 of their badly needed vehicles. It was due to the French loaning 279 heavy transport vehicles, and the RAF taking over 150 from units already evacuated, that the bulk of the personnel of the Air Component and much of the AASF were saved.

The efficiency of AASF operations was gravely impaired during this time owing to the lack of MT. Being under instructions to be prepared to move at short notice, it was impossible to release MT from current sites, and to provision ammunition, accommodation and stores at the new airfields. While only a relatively small number of RAF units were in the evacuation via Dunkirk, the bulk of the RAF was directed to the Cherbourg area. But there was also Haddock Force, of the 71st Wing and supporting units, formed to bomb Italian targets from French bases, which had to make its way out via Marseilles.

Remaining concealed under cover by day and moving in convoy at night, with dimmed lights, drivers in France adopted a marking found useful by British civilian drivers in the black-out. This was the painting of the rear of the differential casing in white, to show up to the driver in convoy behind. The lessons of the campaign were to prove invaluable. In the field, RAF units needed to be fully mobile. Training in convoy discipline was necessary and enforced moves in training would combat the tendency to accumulate surplus and unnecessary equipment.

Vehicles evacuated to the UK were overhauled and reissued. Hundreds were unaccounted for. Many fell into German hands and were used by the German services, some actually being recaptured in 1944. Others went to French ownership, for at Marseilles, owing to the congestion at the harbour, many of the heavy vehicles could not get through to the quayside and were abandoned. Their fate is very obscure.

Western Desert

In the campaigns in the Western Desert, with vast distances to cover, enemy action by shells, bombs and mines, and shortened active life by the wearing effects of sand, there was a heavy attrition rate of MT in this theatre of operations. Deliveries from the UK and, later, direct from America, were shipped in to Egyptian ports to a central MT base pool, for servicing, inspection and delivery to units. At times they were used for transporting supplies en route.

The standard finish for the area was the desert colour of middle stone. As the aircraft camouflage was a disruptive pattern of dark brown and middle stone, the former colour was often daubed over the standard finish in patches. As in France in 1940, unit numbers were not permitted, with the exception of ancillary units not attached to field formations, which bore unit numbers and initials.

Roundels were used on MT in the desert. Theatre orders issued in December 1942 stated that all vehicles, except ambulances, would display an RAF roundel of 31 ins diameter, with a 6 ins yellow outline, on the largest clear upper surface. This was usually marked on the cab, for although the canvas provided the largest area, it was not always fitted. Since regulations were that the small roundel at the front would be placed on the offside, and in Egypt traffic drove on the right, the roundel was normally placed on the left wing—the opposite side to vehicles at home.

2nd Tactical Air Force

Since the invasion force set sail from the UK, the finishes of the vehicles, like those of the aircraft, were the home markings which, as previously related, were

2TAF MT finish as shown by a Crossley Type Q truck with signals cabin, olive drab overall as from September 1945, 'TAF 84' for Tactical Air Force No 84 Group and a chalked number for its convoy position (E.F. Cheesman).

A characteristic of 2TAF vehicles was their large roundel markings—and this one is probably the largest ever displayed on an RAF vehicle, seen on a mobile Ground Control Intercept radar antenna, situated on a Dutch dyke in early 1945 (E.F. Cheesman).

in a state of transition and were to change again to an overall olive green. All vehicles bore a white five-pointed star on cab tops or bonnets, but in addition, those of the RAF had roundels marked large on their side surfaces. RAF units also bore a unit serial number marked on the cab. Apart from heavy and light tenders, utility vehicles and the like, there was a vast number of specialist vehicles and the whole force was fully mobile. The only formation markings carried were the group numbers: 2, 83, 84 or 85.

Far East

In India, the standard blue/grey of RAF vehicles had been maintained. With the campaign in Burma, a general change was made to olive drab. As a roundel marking was circular like the Japanese rising sun marking, a five-point yellow or white star within a 22 ins circle was marked on vehicles in operational areas.

When the war in Europe was over, it was the intention to send a Commonwealth bombing force to the Pacific area, using Lancasters initially and then Lincolns. The first vehicles of Tiger Force had been shipped and were on their way when the Japanese surrendered. These were finished in olive drab with RAF 9 ins transfers marked at the front on the right-hand side, and at the rear of the vehicle. The Force code letter 'Q' appeared in 5 ins letters at the front left, followed by a code number for the unit concerned. The 550 vehicles on the Force establishment did not include any cars, other than utilities, since it was Force policy to use Jeeps and motor cycles for liaison work. Most of 15 cwt vehicles were of Canadian manufacture, and as the RCAF formed part of the Force, so their spares supply would be less distant when the force was in situ on a Pacific island—a stage that was not reached.

Chapter 9

The postwar scene

With the war in Europe over in May 1945, the supply of materials, including a large quantity of vehicles, was switched to India, for use in the campaign in Burma, from where Operation Zipper was to be launched to recapture Singapore. Apart from diverting deliveries from production, there was a surplus held in reserve for Operation Apostle, which could be released. This resulted from contingency plans for the reoccupation of Norway, where it was thought the German forces might hold out.

The Japanese surrender the following August led to the cancellation of new contracts, but for the next year vehicles continued to be delivered off existing contracts, while the Force as a whole started on its rundown to a peacetime establishment. But establishments fluctuated, as the Cold War, together with the Korean War, led to the RAF achieving its peacetime peak in the early 50s, with 6,338 aircraft and 39,500 vehicles on strength. From then onwards establishment and strength figures decreased rapidly to around 2,600 aircraft and under 20,000 vehicles by the end of the 50s.

In the mid-50s the majority of vehicles held by the RAF were still from wartime deliveries, and a few, 235 in fact in January 1955, were pre-war deliveries, comprising 137 tractors, 40 stores freighters, 39 fire tenders, 22 motor cycles, 2 cranes and a car. The purchase of new vehicles was in many cases a saving because of their more economical running and ease of maintenance. Many of the American 3-ton vehicles which had been supplied under Lend/Lease had to be returned and so new purchases were necessary. The severe winter of 1946/47 had found the Force lacking in sufficient snow-clearing equipment and new vehicles were bought for this purpose, but in particular a whole new range of signals vehicles was needed to keep pace with advances leading to the development of new electronic equipment.

Peacetime regulations

An innovation late in the war was the introduction, on an experimental basis, of time recorders in RAF vehicles at home, extended to all RAF prime movers from September 1945. This was the tachograph, which had originated in Germany in 1927. It was explained in a general order to drivers that the recorder was a self-contained precision instrument, which, when fitted to a vehicle, automatically recorded on a wax chart the exact clock time the vehicle was in motion, stationary with the engine running, or stationary without the engine running. This was a measure to ensure full utilisation and economy in running, and one which, outside the Service, some unions have objected to, calling it 'the

Above *In the Near East, cars retained their blue/grey finish but had white tops to reflect heat, as shown by this Akrotiri scene in the 70s. In the foreground are fire extinguisher hand trollies* (K.M. Robertson Collection).

Left *Light warning radar van (RAF160116) exemplifies the changes from war to peace. Its former TAF84 marking has become BAFO84 (under roundel) for No 84 Group of British Air Forces of Occupation. Whereas in the UK and in the British areas of operations only left-hand drive vehicles had 'No Signals' signs here, on the continent, with civil traffic reviving, the vehicle markings to the right of the roundel read 'RIGHT-HAND DRIVE, NO SIGNALS'* (E.F. Cheesman).

Early allocations of the new numbering system for RAF vehicles, show the two methods of display. The Land-Rover has its identity number in three lines as displayed on motor cycles, while the ambulance displays its number in the single line standard method. Above the Land-Rover's number-plate is the formation code, T/23 for No 23 Group of Flying Training Command. This vehicle was based at No 3 Flying Training School, Feltwell (K.M. Robertson Collection).

spy in the cab'. This is in spite of the fact that its use has been in EEC regulations since 1976. However, it was soon realised that it was not practicable to mount recorders in ambulances, fire-crash tenders, motor cycles or armoured fighting vehicles.

The withdrawal of wartime blackout and lighting restrictions brought forth a spate of new regulations, both Service and civilian. As far as the RAF was concerned 36- and 6-Watt bulbs were specified for head and side lights respectively, but 24-Watt bulbs were specified for fog lamps, which were restricted to ambulances, fire tenders and cars of AOCs. Pepperpot and other types of masking were to be removed, but War Department, Home Office and some commercial masks, which could not be modified by widening their apertures, could continue to be used until new rims and glasses were supplied. The height of headlamps from the ground was regulated not to exceed 3 ft 6 ins.

Colour changes

The transition from wartime to peacetime finishes came slowly. It was April 1946 before the decision was taken to cease using paints with gas detection properties and to revert from the olive drab overall camouflage colour to RAF blue/grey in semi-gloss paint, but with valances, wings and chassis in black. As this took effect when vehicles were repainted, it was 1948 before the majority of RAF vehicles were in their new colour scheme. Then in June 1949 came a further change to full gloss in blue/grey. Armoured cars were not affected.

For these new finishes, undercoats of red oxide, lemon zinc chromate and dark battleship grey were used for ferrous metal, aluminium alloy and wood

The RAF Police were privileged with special markings, such as a police board and white bumpers (Peter G. Cooksley).

surfaces respectively. The finishing scheme was two coats of blue/grey enamel (BSC633) and black enamel for the trimmings. For canvas surfaces only, a blue/grey bituminous emulsion was used.

There was a change again in 1954 when special schemes for overseas areas were promulgated as follows:

Area/Command	*Exterior*	*Interior*	*Canvases*
RAF in the UK	Blue/grey (BSC633)	As exterior	Blue/grey
2nd Tactical Air Force	High gloss deep bronze green (BSC224)	As exterior	Olive drab
Middle East Air Force	Matt light stone (BSC361)	Eau-de-nil (BSC216)	Light stone
Far East Air	Blue/grey (BSC633)	As exterior	Blue/grey

Ambulances were exempt from the new scheme and bomb disposal vehicles, while having the overall finish appropriate to their area, had their wings painted in signal red enamel (BSC537). The small roundel transfer was retained for use on the one wing of all vehicles and on bomb disposal unit vehicles it was painted on the opposite side to a 9 ins diameter white disc bearing the letters 'BD' in 4½-ins high letters. Air Ministry Works Department vehicles working on airfields were given golden yellow (BSC356) top surfaces for high visibility, and cab tops in the Middle, Near and Far East were painted white for heat reflection.

During the 70s came the most radical changes of a yellow band around the bodywork of all vehicles liable to work on airfields and then a general change in overall finish to mid-green, but retaining the yellow band.

(Peter G. Cooksley).

MT Unit badge frame, 1946-1952 (Peter G. Cooksley).

MT Unit badge frame since 1952.

Identity markings

From 1949, all service vehicles came under a unified marking system, for individual identification and in lieu of a civil registration plate. This measure, still in effect, consists of two numbers, two letters and a further two numbers. The letters relate to the service, with the Admiralty taking the letters RN and the Air Ministry taking AA to AZ so that the first 'A' denoted 'Air Force' and the second letter could be used as a prefix letter to varying series of numbers. This left a wide range of letters available to the Army.

From August 1946, security regulations were relaxed on unit identities, and a standard badge frame, with the authorised unit badge, could be displayed on vehicles, but no larger than 12 ins high. The standard badge frame was then surmounted by the Imperial Crown, but when Queen Elizabeth came to the throne in 1952, the St Edward's Crown was substituted.

A national distinguishing symbol for NATO forces vehicles was introduced in the 50s. This consisted of a 6 × 4-ins national flag marking, which meant a Union Flag (to give the Union Jack its correct name) transfer, for applying to the front and rear of British Service vehicles on the continent, and for those in the UK serving under direct NATO control.

While the Commonwealth Air Forces provisioned their own transport it was different for Commonwealth squadrons operating under RAF Command. Here a No 408 (RCAF) Squadron Lancaster BII is refuelled by an RAF tender for a raid on Cologne at Linton-on-Ouse in April 1944. The B6 on the tanker's door indicated RAF Bomber Command's No 6 (RCAF) Group (Public Archives of Canada PL-29074).

Part 2
Specialist Vehicles

Chapter 10

Cars, coaches and cycles

Both the RNAS and the RFC had a number of officially allotted staff cars prior to the First World War, and in August 1914, as previously mentioned, a number of cars were impressed and others were presented by patriotic members of the public. In at least one case a chauffeur was thrown in! Thus the cars were of various types, but there was some degree of standardisation since the RNAS concentrated on Clement Talbot touring cars whilst the RFC took almost the whole production of Crossley touring cars. After the war the RAF standardised on the Crossley.

The National Expenditure Committee, during the stringent economy measures to reduce government spending in 1923, was critical of the RAF's 164 touring cars and 34 landaulettes, believing that two landaulettes and 120 cars were sufficient for the whole RAF. Of coaches at that time there were none. As for the carriage of personnel, seats were provided in tenders to a set maximum, as tabled earlier. Two VIP cars were maintained for special occasions. In the UK a Rolls-Royce Silver Ghost was appropriately silver in appearance as well as name. Silver was also the colour of RAF aircraft of the time, due to their powdered aluminium fabric doping. A Crossley tourer in similar finish was held at RAF Middle East Headquarters, Villa Victoria, Cairo.

Cars galore

The RAF expansion in the mid-30s introduced a number of new types to the RAF. An American look was imparted by the semi-streamlined Ford Model 81A Fordor. Other Ford models were followed by Humber Snipes and Super Snipes, Hillman and Standard saloons. During the Second World War over 100 different car models of some 25 different makes were in RAF use. This included a Bentley, used in the Belfast area and a Rolls-Royce limousine at Hawarden, both used by Air Transport Auxiliary (ATA) personnel.

Postwar cars were classified in grades. In the top class, as Grade 1 limousines, were the Austin A135 and Princess and Daimler 4.2 litre, with a few having left-hand drive for overseas service. Grade 2, classed as saloons, included Humber Super Snipes in their third, fourth and fifth series of models from the first of the early war years. There were also a few Rover 3½-litre cars, Daimler Majestics and Jaguar XJ6 models with left-hand drive. Grade 3 comprised Humber Hawks and Ford Zodiac Mark 4s. Largest of the classes was Grade 4 which included Austin 1800s Marks 1 and 2, Ford Executives, Falcons and Zephyrs and Vauxhall Crestas. Some of these were issued in RAF blue/grey and others in glossy black. Finally came Grade 5, which in the early 70s was limited to Morris

Minis. RAF police have had their own choice of cars after evaluation, and in the 70s they were using a Vauxhall Victor police special and Austin Westminster and Wolseley 6/110 models overseas.

Coach comfort

A few single-decker bus types were acquired by the RAF during the war for depot holding, and were used to transport course personnel and special detachments. The first coaches in general use were those introduced for aircrew use at airfields where, as was often the case in wartime, aircraft were widely dispersed. Hampered by flying kit and carrying flasks, maps and navigational instruments, it was too wearing for crews to walk out to their aircraft a mile or so distant. Utility vehicles, 15 cwt and 3-ton trucks were detailed to take the crews out, until special coaches were supplied to RAF stations and ATA ferry pools.

For aircrew transport a special wide coach body was designed to hold 23, with two sliding windows, and side and double end doors to cater for the bulky flying kit. These were mounted on three types of vehicle, the standard Fordson WOT1 6 × 4, the Austin K6 6 × 4 chassis and the 4 × 2 Dodge VK-62B. The Dodges had been ordered in large numbers for the French Air Force. In the same way that the RAF took over French aircraft orders for Curtiss Hawks, Martin 167s and North American T-6s (Harvards), so the MT and material contracts were taken over and Dodges were used as 3-ton trucks as well as personnel carriers. Some of these vehicles with coach bodywork were taken over for recruiting offices. An unusual refinement in these aircrew vehicles was a speaking tube from coach to cab.

Transport Command, taking on the role of an airline after the war, and a

Crossley tourer, Army of Occupation, Germany, 1919 as used by Lieutenant Colonel C.F. Kilner DSO when commanding the 82nd Wing (Group Captain J.D.S. Denholm).

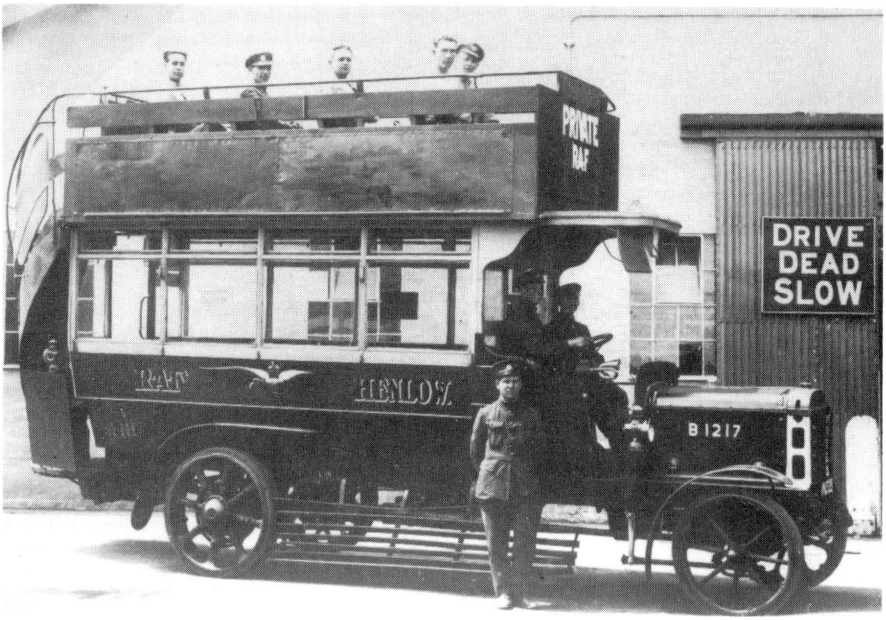

Above *RAF bus in pre-motor coaching days. Used by one of the RAF's largest stations in the 20s, RAF Henlow, the bus, an ex-London General Omnibus Company vehicle, was run for station personnel from non-public funds* (REU8088B).

Below *Bedford 36-seat coach/16-stretcher ambulance, Bedford SB3 chassis with coach-work by Strachans* (Peter G. Cooksley).

large one at that, needed passenger coaches and these were acquired on a grand scale. In 1946 the Air Ministry ordered Commer Q4 Commandos with lower and upper rear compartments, like BOAC coaches of the period. The main service version was the Mark 2 (examples RAF 137900-1) with a Mark 3 version with left-hand drive for overseas. These seated 20 passengers, at a time when the standard passenger transport aircraft, the York C1, carried 24. These vehicles were conspicuously marked on each side with the words 'AIR PASSENGER COACH'.

The Air Ministry issued a colour scheme for air passenger coaches of Transport Command in February 1946, pointing out that the normal regulations governing the identification and marking of RAF MT were being modified to permit the incorporation of the Transport Command crest (the Ministry meant badge) and the name of the appropriate staging post or unit in scroll form. In effect, a livery was being permitted. The colour of the bodywork was specified as fawn grey, with mudguards, valances and wheels in black whilst the radiator shell and bumpers were left in their chromium finish.

Detail markings for the coaches allowed a 20 ins diameter roundel on each side, towards the rear quarter of the body. For this there was a standard transfer; store ref 16B/5532. Another transfer; 16B/5533, referred to the white scroll, lined in black, which was to be placed below the roundel to give the unit or post name or number in 2¾ ins lettering. Yet another transfer (16B/5534) was a magnificent Transport Command badge in full colour, which went on each side of the body between the driver's and the passengers' door. All but the roundel transfers left in stock were declared obsolete and scrapped in August 1967 when Air Support Command took over Transport Command.

A new aircrew coach, a ten-seat 4 × 2 Bedford A3LZ was introduced in the late 50s for transporting aircrew with their new air ventilated flying suits. These coaches towed air-conditioning trailers which were connected to the coach to supply ventilation until the crew linked up with the aircraft system. Another Bedford, the 14-seat Model J1Z2, was introduced in the 60s as a general purpose coach.

The largest RAF coach was a 49-seat left-hand drive Bedford SB3. A similar 43-seat coach using the same chassis type, but with right-hand drive, was made locally in Singapore for the Far East Air Force. Most of the large coaches in service were quickly convertible to ambulances taking 16 stretchers, as part of the war emergency preparedness. One unusual one-off coach type, using a Bedford SB3 chassis, was officially described as a 'Coach, dog demonstration team, adapted with kennels'. At its peacetime peak, in the early 50s, the RAF held 751 coaches.

Motor cycles and bicycles

The bulk of RAF prime movers had four or more wheels with large areas of bodywork for displaying markings, but a motor cycle—basically a bicycle frame with an engine—had only the fuel tank as an area sufficient to carry markings. This was the area used by manufacturers for their trade name or trade mark, and, in some cases, for a maker's distinctive colours or trimming. This puts motor cycles in a class of their own for general service regulations and markings.

Welcome to Gan. Now a staging post of the past, with 3-ton Bedford showing the flag and standing by for baggage, while a Bedford ten-seat aircrew coach awaits passengers. Britannia C1 XL636 Argo *in background* (C.C.H. Cole).

Pre-RAF motor cycles

Motor cycles were a very essential part of RFC organisation. They were required at General, Brigade and Wing Headquarters for carrying despatches to lower formations and at squadron level for reporting operations to higher formations. In army co-operation squadrons the establishment of motor cycles was higher than in other squadrons. A pillion set was provided, but this position was normally used for despatch cases. Three motor cycles was the minimum number required on the road for a squadron move, one with the advance party, one with the main party and one bringing up the rear to move forward and report to the convoy commander any 'fall-out'.

The requisitioned grandstand at Ascot was the main RFC motor cycle store 1915-18. To there, motor cycles were delivered from makers, and from there, issues were made daily for units throughout the United Kingdom, (which then included all of Ireland) and units overseas. To give some idea of their work; in one day (September 27 1916) transactions with a single Wing (18th) were an issue of six P&Ms to No 56 Squadron at London Colney, a P&M with sidecar to Northolt and an exchange of a Triumph for a P&M for Wing HQ, then in the original Adastral House in London. At that time, the RFC was standardising on the Phelan and Moore (P&M) 3½ hp single cylinder Panther and issuing them to replace Triumph and Douglas machines, but when the squadrons arrived in France it was to find that the Army Depots on which they relied for spares catered mainly for Douglas and Triumph machines.

Motor cycle number plates were displayed on top of the front mudguard and on the back of the rear mudguard. As with the rest of the RFC, MT numbers were

Above *A P&M motor cycle in RNAS service wearing a UK registration plate at Mudros in the Aegean, seen in front of a Daimler 3-ton truck conveying Turkish prisoners on airfield levelling tasks, hence the guard with the fixed bayonet in the cab. On the left, a Pierce Arrow truck with RNAS in blue on its khaki bodywork* (via B.F. Hansley).

Below *Norton 1939.* **A**—*speedometer,* **B**—*licence holder,* **C**—*fork spring damper for rough-riding adjustment* (Peter G. Cooksley).

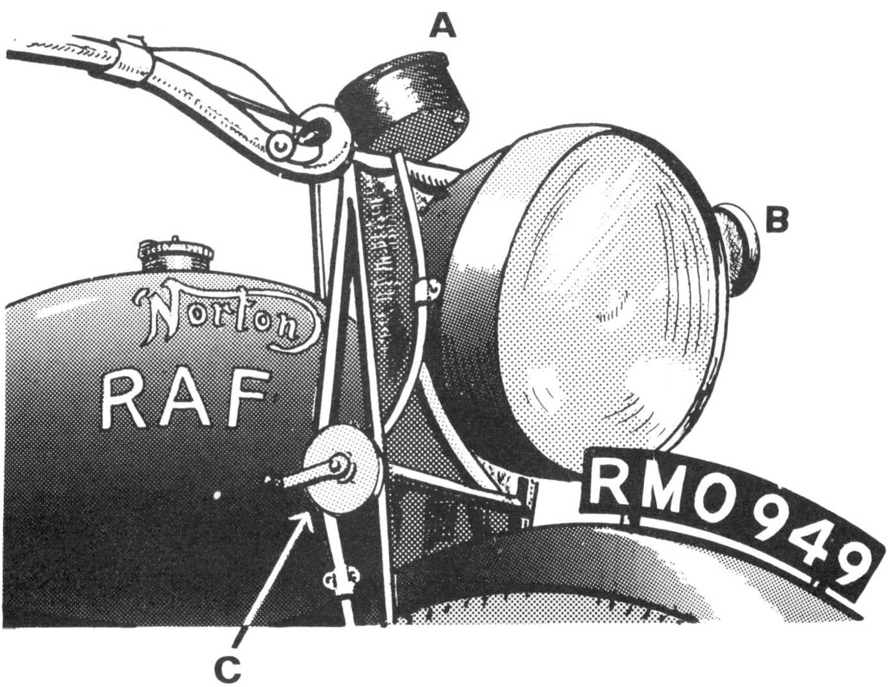

allotted in two separate series. Machines for use with home units and the Air Ministry, formed in 1917, were registered by the local authority with index letters and numbers. Motor cycles for field units had a temporary service number of up to five digits with a 'C' suffix to denote the series. Examples are Douglas 36782C, P&M 40877C and Triumph 36869C.

The RNAS used a variety of motor cycle types, some having been presented early in the war, including the P&M. These were registered by the local authority, and, being their official identity, these number-plates were retained on motor cycles sent overseas.

RAF two- and three-wheelers

By the time the RAF was formed, RFC standardisation on the P&M in the UK was almost complete. In late 1918, at the close of hostilities, the RAF had 3,380 P&Ms, but only 135 were overseas, where the RAF still had call on Army Douglas and Triumph machines, of which some 18,000 of each were in use. Many had sidecars attached and these were known officially as 'combinations'.

A vast disposal scheme got slowly underway. The easily detached Douglas engines were selling boxed, years later, at £5 each. From the Imperial Gift offer to Commonwealth countries, 131 combinations went on to arm part of the

P&M 1918. **A**—*horn,* **B**—*acetylene holder,* **C**—*licence plate,* **D**—*oil pump* (Peter G. Cooksley).

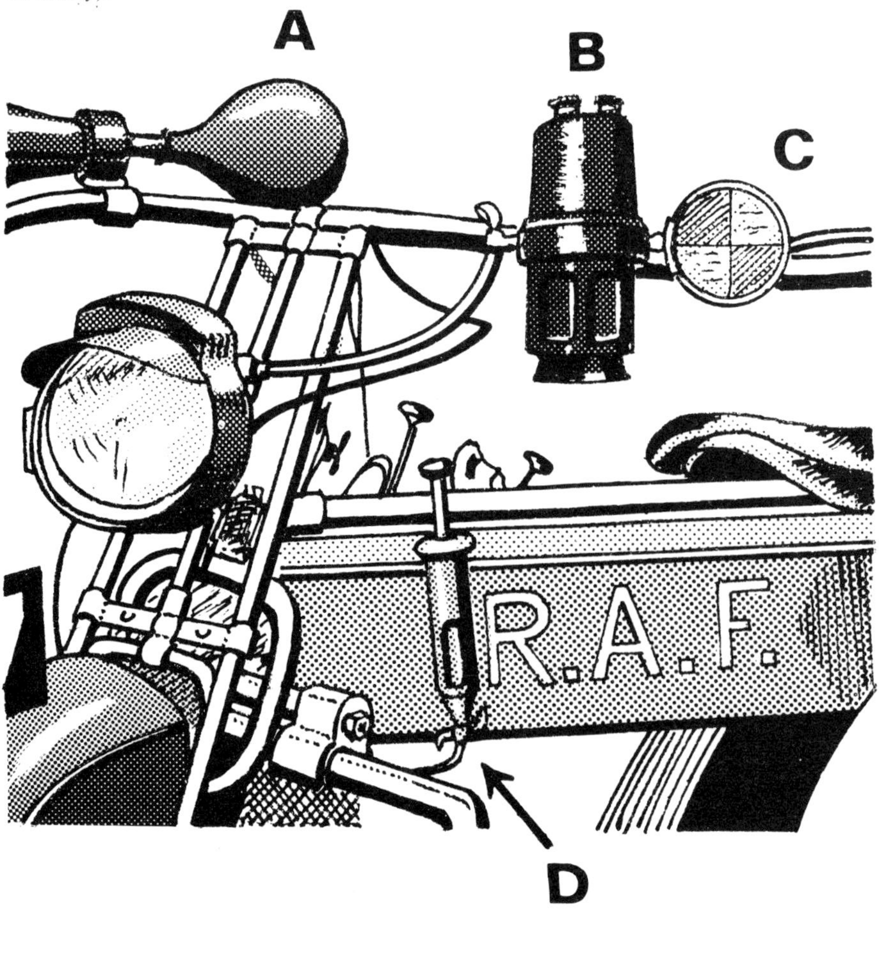

Above *RAF Triumphs photographed during the London-Paris Arch to Arc race July 22 1959. The tanks of these machines bear 1-inch stencilled letters indicating their units. The first has 'TAF/LAAR' for Tactical Air Force (station) Laarbruck and 26AX37 behind has '402' over 'A.S.P.' for No 402 Air Stores Park* (MoD PRB12067).

Below *A P&M motor cycle in RAF service 1918 in the UK. The solid oil tank cap, forward in the tank, can be seen contrasting with the celluloid petrol cap for the petrol tank* (IWM Q12291).

Rear view of RAF Triumph showing grey canvas despatch bags, in this case for carrying film for development from a No 2 Squadron Swift FR5 photographic sortie at Geilenkirchen, July 27 1957 (K.M. Robertson Collection).

initial equipment of Commonwealth air forces. During 1920, 472 combinations remained in use in Britain Egypt and Iraq, while 870 were in store awaiting Disposal Board instructions. At the Cadet College, 100 motor cycles were held surplus to establishment and cadets could sign them out when going on leave. This scheme was stopped after the Treasury had queried how a Cranwell-based machine came to crash 100 or so miles from the station.

The P&M Mark I was formally declared obsolete in 1928 by which time the Marks II-IV were standard. The P&M monopoly came to an end in the late 20s, when the RAF bought OEC Villiers with a 312cc engine. Later BSA, Triumph and Velocette motor cycles came into use and 4.9 hp Nortons with and without sidecars. Most of these saw service in the Second World War together with Indian and Harley-Davidson models from America. Many of these were later sold to the public at around £20 each. Postwar motor cycles in use have included BSA B40, 98cc James 10 and 11 Comet, and Triumph TRW. The numbers held have not been revealed, but the latest released information in Public Records shows that in the early 50s around 1,354 motor cycles were in use with a mean age of five years and that in RAF service they had a ten-year life expectancy. However, over 20 in use had been on the road for 15 years.

The antithesis of camouflage, RAF Police Triumph motor cycles in white for high visibility, at a field exercise in March 1978. It will be noted that number-plates are no longer attached to front mudguards, for safety reasons (MoD Pub PRB5503/45).

RAF motor cycle marking and colouring decrees

With little surface area to display RAF ownership, the first Air Ministry order on marking motor cycles advocated the letters R.A.F. (it was invariably with stops then) being marked on the top bar of the frame. In 1922 instructions were issued to mark RAF on the tank in the manner illustrated, which some units had done since 1918. After the war, in the UK, the motor cycles held by the RAF were identified by their civil licencing numbers, regulated by the Road Vehicle Regulations. Registrations were displayed on front and rear number-plates in white on a black background. Later they came within the RAF vehicle numbering system and in the 50s into the service numbering series of two numbers, then two letters (of which the first is 'A' for Air Force) and a further two numbers.

From 1939, as with other vehicles, motor cycle headlights were shaded to comply with the blackout regulations, but because of the lack of bodywork, camouflage paint was not considered necessary. It was different for the riders. Crash helmets introduced early in 1941 had a shiny finish. To avoid the danger that this headwear would compromise camouflage, instructions were issued later in the year to paint them Matt Camouflage Green No 3, in contrast to RAF

Bicycle maintenance at Binbrook mid-1944. It will be seen that rear mudguards are painted for 6 ins in white in accordance with the civil road regulations during the blackout period (E.D. Evans via Garbett/Goulding Collection).

steel helmets, which were painted grey. After the war, when road safety became of overriding importance, white bands were introduced and later all-white helmets.

Bicycles

Do not despise the bicycle, for it was the most widely used RAF vehicle, and particularly so in the Second World War, when domestic sites were often miles from technical sites or aircraft dispersal. The RFC classed bicycles in two types; standard pattern army bicycles which in the RAF became known as 'Service Pattern' bicycles and those bought locally, known collectively as 'Trade Pattern' bicycles. This distinction was made because spares for the former were supplied by No 1 Stores Depot at Kidbrooke and for the latter by local purchase. An allowance per bicycle, per annum, was laid down by the RAF in 1918 as: 1 pint general service lubricating oil, 1 lb tallow, 1 oz cotton wool (for replenishing lamps), 1 bicycle brush, 1½ lbs of old linen, and 1 quart of paraffin for cleaning and replenishing lamps and wicks as necessary.

There was rigid economy in the supply of bicycles between the wars, ending in 1940 with the need for wide dispersal. By then there was a further classification by sex, Bicycles Male and Bicycles Female, as the WAAF had occasion to use bicycles officially. Bicycles were normally issued to sections of a base, rather than to individuals, and so were often marked with flight or other detail on their mudguards; additionally, of course, many private bicycles were used and their use was encouraged.

Chapter 11

Fire, crash and rescue vehicles and ambulances

In the early days of the Royal Flying Corps, fire prevention services were supplied by the Army Service Corps, but from 1916, the RFC was operating its own fire engines at home flying stations, in co-operation with local authorities. Their general guide on fire-fighting principles was the *Army Fire Manual* of 1917. This concerned mainly preventative measures and was rigidly followed, particularly overseas in the First World War, when fire engines were not generally available for flying units in the field. The policy of allotting the available fire-fighting mobile equipment to units at home, rather than those operational overseas, was logical in the light of flying intensity. Training units at home would be continuously flying throughout the day, whereas, except in critical periods, operational patrols or bombing missions per airfield at the front would rarely average more than 24 flights per day.

The lack of suitable fire tenders was brought home in 1919, when a series of fires resulted in the loss of over a hundred aircraft. Such a deficiency was brought out by the Court of Inquiry on the loss of Handley Page 0/400 D8314 and its giant hangar at Ligescourt on May 21 1919. The cause of the fire, which was in the early afternoon, was not properly established, though it was believed to have been caused by the heat of the sun on the hangar camouflage screening. Fire Standing Orders for No 100 Squadron stated that, on the fire alarm being sounded, drivers would stand by their vehicles to drive them away from the danger area. This was carried out and all the available fire-fighting equipment (36 Minimax and 56 Pyrene extinguishers and 24 sand-boxes) was used but the hangar could not be saved and the tanks of the aircraft exploded. Only the fact that it was the end hangar, with the wind blowing the flames away from the others, prevented the loss of more of the squadron's aircraft. It was brought home that a mobile fire engine with hosing was essential for each station, even under the economies of peacetime. This was in addition to the ambulances, which, it should be appreciated, were then also the crash vehicles of the day.

As an interim measure, Crossley tenders were modified at the MT repair depot during 1921 to carry 30-gallon chemical extinguishers and Fire Snow hand extinguishers. The following year an RAF Fire Fighting School was organised on a temporary basis at Cranwell, with instruction being provided by the London Fire Brigade. During the mid-20s a number of civilian type fire engines were acquired from Merryweather and, by the 30s, standard tenders such as the Crossley P and Morris B were being ordered with fire-fighting bodywork on the standard chassis. These were the standard fire fighting vehicles of the Force when war came in 1939.

Merryweather fire engine seen at a Royal Flying Corps training station in 1917 (E.F. Cheesman).

A frequent point of contention was the fact that, for any fire at an RAF station, the local civil fire brigade could be called in, but for civil or municipal property on fire near an aerodrome, the RAF fire engines would not attend. This was because the local authorities were charged with suppressing all fires in their area, while an RAF fire tender, according to regulations, could not leave its area of duty.

Fire fighting in the Second World War

Following the RAF expansion, additional orders had been placed for improved models of standard fire tender types for the new stations which were opening. Such vehicles were normally on a station equipment, not on a unit, so that squadrons did not have a fire-fighting vehicle on strength. For mobile units, protection was available from fire-fighting trailers, similar to those used in large numbers by the civil Auxiliary Fire Service.

In 1940, Air Ministry Fire Services instituted a new approach to fire prevention, because of the large stocks of inflammable material which were piling up at sites all over the United Kingdom. Large orders were placed and existing contracts extended for fire, crash and combined fire/crash tenders. Standard Austin K2 and Karrier chassis were fitted with domestic type fire-fighting bodies. Also under Lend/Lease, Chevrolet 1½-ton 4 × 4 fire/crash tenders were delivered to the RAF.

Fire-fighting equipment was particularly lacking in the Middle East. Here the RAF was dependent on the Army Fire Fighting Equipment Pool for equipping

Above *Merryweather domestic type fire engine with escape ladders, being groomed at Hendon after being brought down from Cranwell to supplement Hendon's Crossley fire tender during the annual RAF Display* (MoD H498).

Below *Bedford 3-ton general service chassis adapted as fire tender. This example was photographed still in service at Wattisham in 1953* (K.M. Robertson Collection).

Above *A look at the works of the Mark 5 Thornycroft (45AF44) showing the bodywork containing foam and water compartments, control points and lockers* (K.M. Robertson Collection).

Below *A Mark 6 with monitor in action, with support from a Mark 7 coming up on the right, during a practice at RAF Nicosia May 29 1962.*

base units, so their fire tenders bore identity letters prefixed 'WD' for War Department; not 'RAF', whose charge was indicated by a small roundel. Most of their vehicles were in khaki paint as a desert camouflage. The Greek Government had ordered a number of Federal fire tenders shortly before their country was invaded. These vehicles eventually found their way to the Army pool which allotted them to the RAF for stationing at their main Egyptian bases. As they proved their worth, 150 were sought from America by direct purchase and eventually 100 of International type reached the RAF in the Middle East under Lend/Lease. These were similar tenders carrying a 680-gallon water tank with a 300-gallon per minute centrifugal pump forward, driven by the road engine.

The School of Fire Fighting had been moved from Cranwell to Weeton in July 1940, and to Sutton-on-Hull on August 18 1943, as the School of Fire Fighting and Anti-Gas. A No 2 School, opened at Ismailia, was closed after the war.

The main fire-engines in RAF service during the Second World War, or ordered during the war and used in the immediate postwar years were:

Chassis	*Wheels*	*Introduced*	*Extinguishing media*
Crossley FE1	6 × 4	1938	Foam & carbon dioxide
Crossley FWD	4 × 4	1940	Foam & carbon dioxide
Fordson WOT1	6 × 4	1942	Foam only
Fordson WOT1A	6 × 4	1945	Foam & carbon dioxide
Jeep	4 × 4	1945	Crash rescue equipment only
Austin K6	6 × 4	1946	Carbon dioxide only
Dodge	4 × 2	1946	Reserve water only
Bedford OYC	4 × 2	1947	Reserve water only

A Mark 6 Alvis/Pyrene fire tender with monitor forward and searchlight at rear, seen at Akrotiri circa 1960 (Pyrene 8054).

The Fordson WO1s were an emergency expedient during an acute shortage of fire-fighting equipment in the mid-war years and some 350 were produced, of which nearly half were modified to make them more efficient in 1944. The WO1A model had a strengthened chassis and again some 350 were acquired. In 1947 the Dodges went out and the main types left in service were retrospectively designated as fire tenders, Marks 1-4.

For some years after the war, the RAF was conditioned to use wartime equipment. Much experimental work was carried out and many extinguishing methods were tried and rejected for various reasons such as cost, weight or toxicity. Foam, by then adopted worldwide, was still seen as the best medium because of its smothering, cooling and non-toxic characteristics. During the 50s, the RAF Mark 5 fire crash tender was introduced to replace the 1944 wartime-pattern tenders then still in service. These new 10-ton vehicles, capable of 60 mph, were nearly twice as powerful as their predecessors, producing 2,300 gallons of foam against the formers' 1,400 gallons. By replenishing from 500-gallon water tankers, a single tender could produce 5,000 gallons of foam.

In the late 50s, the RAF set out to obtain a general purpose crash tender, capable of holding 5,000 gallons of fire extinguishing foam, which would combine high mobility with simple operation and ease of maintenance. This resulted in their purchasing the Mark 6 GP Alvis Salamander crash truck, based

The latest Mark 10 Leyland Nubian fire tender coming into RAF service in the 80s (Scammell C1594/3).

on the Alvis Saracen armoured vehicle and powered, like the Mark 5, by a Rolls-Royce engine. This was followed by the Mark 7 from Thornycroft with an automatic gearbox, a Mark 8 3-ton 4 × 4 Bedford RLH23 with pumping equipment driven by a 4.2-litre Jaguar engine and the Mark 9, carrying 1,250 gallons, of which 77 were acquired. Now the Mark 10 is reaching the Force in the 80s.

Supplementing the foam-producing tenders over the last 20 years have been Land-Rovers for immediate rescue work. One of the first to enter service was designed by Foamite Ltd, using a long wheel base Land-Rover which was capable of 60 mph on its own and was designed to tow a two-wheel trailer of up to 1 ton. Included in its equipment was a long range searchlight and an extendible site light for all-round illumination. In the 70s, different types of Land-Rovers were in use for domestic, tactical crash and airfield crash rescue use. The latest types, classed as Trucks, Aircraft Crash Rescue, are the TARC 1, of 100 gallon capacity, based on a long wheel base Land-Rover and a TACR2 of 200 gallon capacity based on a Range Rover.

During the 70s the RAF used 2-, 4- and 7-ton Bedfords for domestic fire-fighting. A single amphibious fire-fighting vehicle was also in use, being a converted Alvis Stalwart Mark 2. Coventry Climax fire pump trailers, later versions of those used in the Second World War, continued in use. Now, in the

30 AG 98

Left *El Adem Desert Rescue Team in the Near East in the early 60s with a Bedford 3-ton truck (65AF68) and Land-Rovers* (K.M. Robertson Collection).

Below *Gloster Saro fast response vehicles (30AG98 in foreground), based on the Range Rover chassis, in green bearing yellow airfield bands* (Gloster Saro Ltd).

80s, the conventional red of fire vehicles has given way to the standard overall finish of mid-green with a yellow band around the vehicle. Only at St Athan were fire engines still in their traditional red, as of 1982.

Rescue services

Between the wars, RAF armoured cars were frequently employed in search and rescue in conjunction with aircraft, particularly in Iraq. During the war, with good army/air co-operation, it was the army who did rescue work, and formations like the Long Range Desert Group were given details of aircraft which failed to return. After the war, in the Middle East, the RAF maintained a desert rescue team based on Fayid. Team equipment was an 8-ton Matador with a red and white diced body for high visibility. It was equipped with wireless, first aid equipment and 40-gallon tanks of drinking water. To widen the search area, a desert duty Jeep (RAF123953) was attached, using the Matador as a base. The Jeep was later replaced by a Land-Rover.

An RAF mountain rescue service was started in 1942 by Flight Lieutenant F.W. Graham at RAF Llandwrog. With crashes in the mountains occurring at not infrequent intervals, Jeeps and Humber cars were made available to the team. Their first survivor in July 1943 was an Oxford trainee pilot who belly-landed at Tal y Cafn and was picked up, suffering from shock, at the local police station. A month later, following reports of a crash in the Carneddau mountains, a crew of five was picked up, shaken and bruised, 3,000 feet up on Foel Fras. By the end of the year, 33 survivors had been rescued from 22 crashes.

In 1944 the Air Ministry took over responsibility. The original team was put on a proper footing and teams were established at three other stations. After the war there was little activity, but new teams were established with a small permanent staff and up to 30 volunteers. By 1950 there were nine teams and the original team had moved to Valley, on Anglesey. The MT available to each was a 3-ton lorry permanently loaded, call on two Humbers, one for signals and the other an ambulance, and Jeeps. Currently the vehicles are Bedford trucks and Land-Rovers.

An even more ambitious rescue organisation was started in Cyprus, at Near East Air Force HQ, for operations in their NATO area. For operations in adjoining countries of the alliance, all members of the team were trained as parachutists, with Land-Rovers which could be parachuted down for land search and rescue. Much of their work, and that of the mountain rescue teams, has now been taken over by helicopters, but there is still a need for such services by MT in certain areas, in adverse weather conditions, and as a supplement to air search.

Ambulances and medical trucks

At an international convention of European Powers at Geneva in 1864 there was agreement on humane regulations for the treatment of wounded in warfare. It was agreed that the display of a red cross on a white background would denote hospitals, aid posts and transports for the injured and wounded. From that time, British Army horse-drawn ambulances displayed the required marking which now has broad acceptance under the United Nations. The marking is normally taken to be a reversal of the colours of the Swiss flag. However, since this is the Cross of Christendom, Moslem countries adopt a Red Crescent in lieu of a cross.

Above *Mountain Rescue Teams in a joint exercise in early post-war days, led by a Jeep (RAF208343) with trailer* (K.M. Robertson Collection).

Below *Latter-day mountain rescue team; a Bedford 3-ton truck with trailer and long wheelbase Land-Rovers (13AA68 ambulance and 85AA49 carrier) and standard Land-Rover (14AA03)* (MoD TN6438/66).

In the early days, the Royal Flying Corps, both at home and in France, was served by the ambulances of the Royal Army Medical Corps and it was not until the time the RAF was formed that it was allotted ambulances of its own for flying fields, with the transfer of 612 officers and men to the RAF from the RAMC, and the attachment of Army medical officers. The first of the RAF's ambulances in general use was the Ford Model T. It was in khaki finish with the red cross on a circular white field. In the early days of peace, the RAF had an ambulance body to suit their Crossley chassis. Later the Albion Type AM463 and two types of Morris chassis were used with ambulance bodies.

At home, blue/grey was the standard colour between the wars, with the Red Cross on a circular field. Overseas, ambulances were overall white by an order of November 1937. During the war, several ambulances came into use, including the Standard 4 × 2 light ambulance for local use and the Fordson WOT1 6 × 4 heavy ambulances. For most of the war, ambulances were camouflaged according to period and theatre and red crosses were of reduced size. In the Western Desert, squadrons were at one period ordered to prepare one vehicle for medical supplies and as a medical aid post, qualifying them for Red Cross protection, but as the vehicles remained in the proximity of the rest of the camouflaged squadron transport they were not normally marked with red crosses.

The Fordson ambulances, praised for their work in the Western Desert, were found cumbersome for operating in country lanes on the continent in 1944. This led to the RAF seeking Army Austin K2 and particularly Chevrolet ambulances.

After the war the RAF had a surplus of ambulances, and ten years after the war, over 800 ambulances were held against an establishment for 700. The Fordson Thames Sussex 6 × 4 and Austin K2 4 × 4 chassis ambulances, which took four stretchers and saw war service, were still available in the 70s, when the main range of ambulances consisted of types of Bedford 4 × 2 ambulances. Currently, as a reserve, there are coaches which can be used as ambulances, taking 16 stretchers each. Also, over the years, the RAF has used many specialist medical vehicles such as the Bedford SB1 5-ton 4 × 2 mobile medical X-ray truck and Bedford 1-ton 4 × 2 renal unit. Unit badges and roundels have been displayed on RAF ambulances and the vehicles of medical establishments. The units concerned and the nature of their badges were, and still are in some cases:

Unit	*Badge description*
RAF Hospital Cosford (now disbanded)	Avenue of trees representing the road to health
RAF Hospital Ely	Cross and three ducal coronets (crowns of See of Ely)
RAF Hospital Halton	Cross and lamp (symbolic of Florence Nightingale)
RAF Hospital Nocton Hall	Lincoln Imp and Rod of Aesculapius
RAF Hospital Uxbridge	Sword and Rod of Aesculapius
RAF Hospital Wroughton	Cross and castle of Swindon's armorial bearings
Central Medical Establishment	Pair of scales and entwined serpent
Medical Training Establishment and Depot	A torch entwined by a serpent

Above *Re-organising for land running after a Land-Rover air drop of the NEAF (Near East Air Force) Para Rescue Team* (NEAF Publicity 2330).

Below *A Fordson 6 × 4 ambulance on the left (47AA93) marked C/19 for Coastal Command's No 19 Group, and an Austin 4 × 4 ambulance taking casualties from a Comet C2 diverted from Lyneham to St Mawgan due to fog, November 28 1958, with patients from Cyprus. Both ambulances are in blue/grey finish with white interiors* (K.M. Robertson Collection).

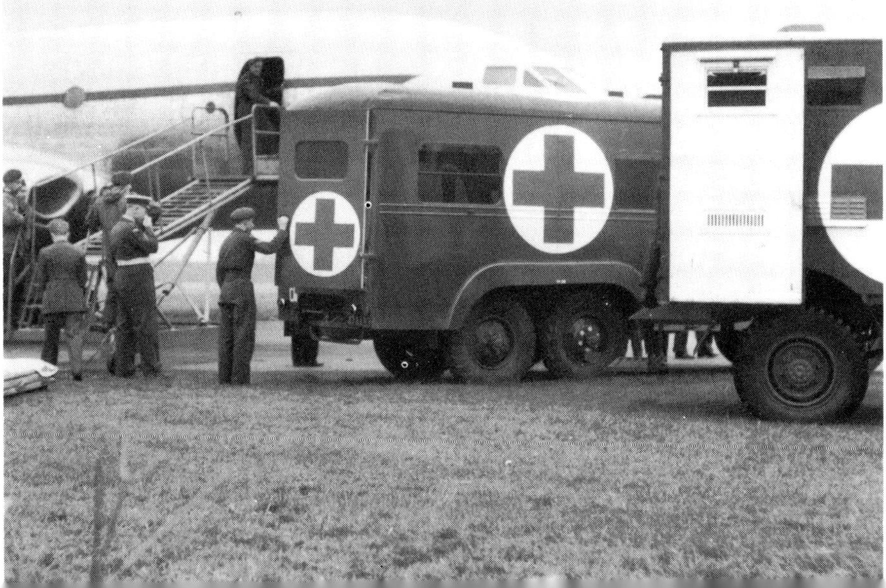

Above *All-white Morris 6 × 4 type B ambulance at Aden; a Vickers Vincent in background. Military vehicles did not require civil registration in the colony. The number-plate is marked 8SQN3 for No 3 vehicle of No 8 Squadron, the sole flying unit at Aden in the mid-30s* (Flightlines E-89).

Below *All-white Bedford ambulance with front wording reversed to reflect the right way round in driving mirrors* (Peter G. Gooksley).

Mobile dental units

The RAF Dental Branch was formed in 1930. Hitherto, RAF personnel had been treated by seconded Army personnel. Dental units were established at RAF hospitals and centres, precluding any need for mobility. Airmen went to the dentist, not the dentist to the airmen. In 1940 there arose the problem of dental treatment for the scattered RAF units at home and overseas. This grew acute, and mobile dental surgeries became a necessity.

As a preliminary step, it was hoped to acquire dental caravans from local educational authorities in Britain, who ran just such a service for schools in isolated rural areas, and 18 were ordered. However, for overseas, shipping could not be spared for such large and bulky items. In the Middle East, utilising such vehicles as could be made available, a Cairo coach body builder was found to supply pairs of dental surgeries and dental laboratories for making false teeth, sufficient to accommodate a standard field dental outfit pack-up, manned by an officer, a dental mechanic, a driver and orderly. In some cases, one of the twinned units was a trailer. Four mobile dental units served with the Desert Air Force in North Africa and Italy and two in other parts of the Middle East.

At home, the ex-school type caravans were used, hauled by Humber Snipe saloon cars modified for towing. For the invasion of the continent, 16 dental surgeries were required, by which time three types of mobile dental units were in use, using modified Albion and Fordson ambulances. Mobile dental trucks have remained on the Force's inventory in recent years, with Bedford RL 3-ton 4 × 4 trucks carrying a dental clinic and dental laboratory.

Chapter 12

MT for fuel and fluids

The British Services during the First World War carried fuel in the standard screw-capped, metal-handled, two-gallon petrol cans which littered France in their millions in 1918, or in bulk in 50-gallon barrels. Standard 3-ton heavy tenders could take 13 barrels or 350 cans per vehicle. For delivery to railheads, a standard four-wheel railway truck could take 875 cans properly stacked, yielding 1,750 gallons. Some Leyland 3-ton chassis were fitted with large tanks for bulk delivery. For aviation use, high grade spirit was used, as allotted by the Petroleum Board. Much of it was from Asiatic sources, until around the time the RAF came into being, in April 1918, when shipping difficulties cut off normal supplies. The RAF at home and in France was then dependent upon American FGAS (Fighting Grade Aviation Spirit) which was tinted red for identification.

For the permanent peace-time aerodromes, a fuel dump was established at each airfield, well away from hangars and domestic sites. Delivery to the station dump was by the commercial vehicles of the oil companies, initially the British Petroleum Company (BP) and the Anglo-American Oil Company. After the RAF Expansion in the mid-30s, the fuel suppliers were, proportionally, Shell Mex and BP 48 per cent, Anglo-American 28 per cent, Trinidad Leaseholds 20 per cent and Corles, Capel and Leonard 4 per cent.

Refuellers

Initially, the tanks of RFC and RNAS aircraft were filled by pouring from the 2-gallon cans. This often necessitated using a ladder and two-wheel tray-like trolleys to carry the cans around. The average single-engined two-seat aircraft of 1918 rarely had a fuel capacity beyond 50 gallons, but the Handley Page bombers then in service in numbers had a total tankage of 370 gallons each, which needed large supply tanks and some way of hosing the fuel into the tanks under pressure. This was met by adapting the standard trolleys for carrying camp water-tanks to carry a large fuel drum with hand pump. These trolleys were then issued generally to flying stations.

During the 20s, the RAF placed orders for prime mover tankers which would use the chassis of the vehicles in service in other forms. By 1939, three main types of refueller were in use; the four-wheel Albion AM463, the six-wheel

Opposite *First World War Handley Page bomber refuelling progression. A hand pump fitted to fuel tanks on a hand trolley, on a trailer and on a prime mover* (IWM Q11534 and E.F. Cheesman).

KEEP CLEAR OF THE
PROPELLERS

P 9187

Albion tanker of the 30s (Peter G. Cooksley).

Morris YI and the Karrier C. For the supply of its own MT in convoy, there was a 450-gallon petrol RAF Type J trailer.

The RAF expansion of the late 30s brought increased orders for refuellers. The Battle of Britain brought out particularly the need for speedy refuelling for fighter turn-round. Controllers' orders to formations to 'Pancake' had a precise meaning—'Land, re-arm and re-fuel' so as to remain at immediate readiness. In the following bombing campaign, with 1,000 bombers operating, many of them four-engined, fuel supply and delivery were severely taxed. Some 24 bomber stations would consume 24,000 gallons of fuel per hour at peak sortie periods. For a complete 1,000 bomber operation, 2,000,000 gallons of petrol, 70,000 gallons of oil, and 5,000 gallons of coolant were needed for the aircraft—and 8,000 pints of coffee for the aircrew!

The wartime tanker, with its long arms which lifted hosing clear for the nozzle to drop into filler apertures, was seen as the ideal delivery system, but it had its drawbacks. Fuel pumping rates of the war period were around 30-50 Imperial gallons per minute. Without opening and closing caps, it took over 1½ hours to refuel a Lancaster to its maximum of 2,154 gallons. With a station base of two bomber squadrons, operating at maximum effort of 40 Lancasters, it will be seen that to refuel the force in eight hours would take five tankers. Therefore, overall, the RAF had a need for thousands of tanker-type vehicles.

Rigid safety precautions were always observed when refuelling. Both vehicle and aircraft had to be properly bonded and earthed to prevent a surge of static electricity causing a spark. Aircraft earthing was usually achieved through its tailwheel of metal-impregnated rubber. The vehicle was earthed by a chain which it dragged along the ground but, as an additional precaution before

The drawbacks involved in over-wing refuelling—working on the wings with hoses dangling heavy metal nozzles, which on occasion caused damage to the aircraft. Here an AEC tanker in postwar blue/grey high gloss, with yellow upper-works and doors for high visibility, is about to refuel a Meteor NF14 at North Luffenham (K.M. Robertson Collection).

refuelling, an earthing rod was driven into the ground. Wartime regulations for refuelling were that the bowser should be headed in such a position as to permit a quick drive-away in case of emergency, and trailer tankers in use were normally kept attached to their prime mover for the same reason. It need hardly be said that no smoking or naked lights were allowed within 50 yds of refuelling operations.

After the war there came a move away from over-wing refuelling. It was a lengthy process requiring several men or women, at times with access ladders, facing the dangers of slipping on icy surfaces or in high winds. The open apertures were liable to let in dust or rain, contaminating the fuel, and there was also a health hazard from fumes.

The modern pressure refuellers cut down the process to a matter of minutes by discharging up to 500 gallons per minute to underwing fuel points, where handling crews work at ground level and connect hoses to aircraft with couplings of agreed NATO standard. By the early 70s, the RAF had standardised on Leyland and AEC Mammoth 6 × 4 10-ton pressure refuellers of 2,500-3,520 gallons capacity, although 3- and 7-ton 4 × 2 Bedford refuellers of 1,000 gallons capacity were also in use. By this time refuellers were officially classed as 'aircraft fuel servicing trucks'.

Tanker/refueller/fuel servicing truck markings

While tanker overall finish has compiled with the regulations of vehicles in general, additional special markings have been used. With metal tanks, the standard RAF letter plates for screwing on to the wooden bodywork of vehicles could not be attached and the letters 'RAF' were therefore painted on tank sides

PMP 362
825

Right *The new era of markings for tankers as introduced in the late 50s* (Peter G. Cooksley).

Left *Joint refuelling and arming of a Wellington, with tractor-drawn tanker trailer* (Real Photographs 825).

Below *Albion AM463 tanker, showing its Zwicky pumping equipment, servicing Gladiators in Egypt, circa 1940* (Zwicky CSR8983).

in white. In September 1935, an Air Ministry instruction decreed that all tankers and refuellers would be used exclusively for one type of fuel and that they would be permanently marked to that effect. Fuel specifications were given a number by the Directorate of Technical Development of the Air Ministry, the two fuels then in use being DTD Specifications 224 & 230. These numbers were marked on the appropriate tanks.

Now came an object lesson for those who decree which markings shall be used; it is a matter which requires some forethought. The original instructions on fuel specification marking stated that within an 18-ins diameter circle, '224' would appear in blue and '230' in red. But it so happened that, to distinguish visibly between the two fuels, the Specification 230 was tinted blue. This meant that the red indication would be misleading. As a result, in January 1936, the order of the colours was reversed in an official instruction. Yet again there was a snag. Since, as related, the standard finish of the period was an overall blue/grey, the blue marking was barely discernible. So the following October a new instruction put the numbers in white, with an additional sign, as 100-octane fuel was then coming into use.

During the war, when these fuel markings were seen to compromise camouflage, they were toned down. After the war, during the 50s, under NATO standardisation, new fuel specification numbers and terms came into use as follows:

AVCAT Aviation fuel for shipborne use
AVGAS Aviation gasoline for piston engines
AVPIN Starter fuel for turbine engines
AVTAG Aviation fuel with an additive
AVTUR Aviation turbine fuel

The new specification numbers for the various liquids carried in tanks were given utilisation prefix letters, using indicative letters as follows: C—Corrosive preventatives, F—Fuel of various grades, G—Grease, H—Hydraulic fluid, O—Oil of various grades, S—Special fluids. These letters appear as prefixes, with hyphen, to a NATO specification number for the fluid.

On various tanks, other significant letters may be seen, with meanings as given: AL—miscellaneous fluids (eg, anti-freeze), OC—Oil compound, OEP—Oil Extreme Pressure (such as gear tooth pressure), OF—Oil Fatty, OM—Oil Mineral, OMD—Oil Mineral Detergent, OX—Oils miscellaneous, LG—Lime-based grease, XG—miscellaneous greases, ZN—special, PX—Protectives miscellaneous.

Tanker fuel markings of the late 30s (Peter G. Cooksley).

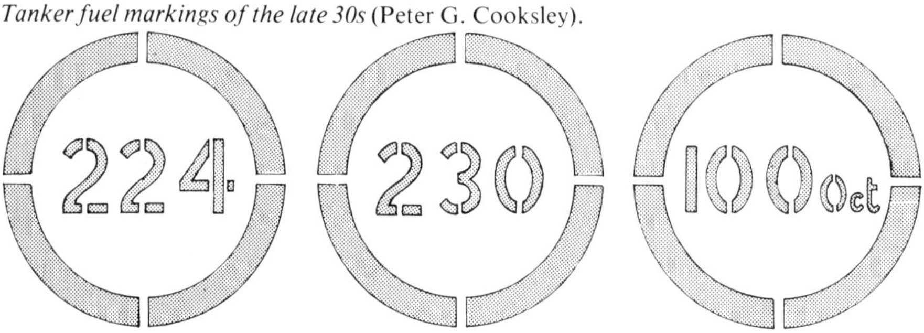

Above *A view of the complicated pumping controls of a modern fuel servicing truck* (Peter G. Cooksley).

Below *For safety for units at home, an overall high visibility yellow was introduced for tankers because of their proximity to aircraft. This is exemplified by a Bedford fuel servicing truck* (Peter G. Cooksley).

Above *1982 operational turnround servicing of a Lightning F6 of No 5 Squadron at Binbrook with AEC tanker, in the current green finish with yellow airfield band, dispensing AVTUR* (P.H.T. Green).

The US Air Force has used, and still uses, jet fuel designations of their own which are also widely used commercially. Since some are equivalent to NATO standards as marked on RAF vehicles, these fuel designations are included here; JP-1 USAF specification, JP-2 and 3 now obsolete, JP-4 equals AVTAG FS11 (NATO F-40), JP-5 equals AVCAT FS11 (NATO F-44), JP-6 now obsolete, JP-7 commercial, JP-8 equals AVTUR FS11 (NATO F-34), JP-9 and 10 commercial, JP-TS fuel of high thermal stability.

Fuel supply

For most RAF stations, the fuel supply remains in the 80s the same as it was in the 20s, supplied by fuel company vehicles. However a few of the major stations are now supplied by pipeline and, up to time of writing, one station in Scotland still receives its fuel by rail.

Field supply and refuelling are vastly different from those at stations. Fuel can be stored in large air transportable rubberised containers, which themselves are wheel-shaped for literally rolling into position. Giant pillows holding up to 10,000 gallons each can be air dropped, and from these the Harrier force can be fuelled during operations. It just needs a pumping unit to provide a fuel transfer and this can be met by a TART (Tactical Aircraft Refuelling Trailer) which combines a pump, water filter separator and hoses, and can be towed by a Land-Rover.

Plain tanker

Apart from the refuellers, or 'fuel servicing trucks' in NATO parlance, there was and still is a place for the plain tanker, or bulk liquid carrier, for oil or special fluids. Water too has always been a primary requirement of any force. In the immediate pre-war years, the RAF had four types of water tank trailers in use; types A, B and J, and the Mark VI, which had a design lineage beginning before the First World War.

During the 70s the RAF had Austin K9 4 × 4 1-ton water tankers of 200 gallons and Bedford 4 × 4 3-ton tankers of 600 gallons capacity. There was also a 7-ton 4 × 2 Bedford of 600 gallons capacity for aircraft water replenishment, and a similar Bedford as a water methanol carrier of 2,950 gallons. For field use, two-wheel water tank trailers are still in use, such as Sankey ½- and 1-ton trailers of 100 and 200 gallons capacity respectively.

Chapter 13

Mobile cranes, lifts, aircraft tractors and agricultural vehicles

Cranes were not provided in the field for the Royal Flying Corps, who were dependent on the Royal Engineers for heavy lifting. For lifting in an engine change or when salvaging a force-landed aircraft, three men could manhandle a rotary engine of a single-seat aircraft. This weighed around 400 lb dry. Most of the stationary engines weighed rather more, and so sheer legs, pulleys and chains, which could be carried in Leyland 3-ton tenders were supplied to squadron level.

A need for cranes at aircraft depots became evident and these were supplied in commercial finishes. Where cranes were installed overhead, mounted on walls, or were embedded in foundations, they came under the Air Ministry Directorate of Works and Buildings and their commercial finishes were retained. Mobile cranes became RAF MT and so were given khaki finishes until the late 20s when blue/grey became the overall colour. Currently those serving on airfields are painted a high-visibility yellow.

The two main suppliers of cranes to the RAF have been Coles and Ransomes & Rapier. One of the first standard types was the Albion AM463 fitted with an electric revolving crane. During the Second World War, the RAF used two main types. The AEC 0854 6 × 6 chassis, illustrated previously with a refueller body, was adapted to take a turntable with a Coles Mark VII crane. In all, 192 were built, mainly for the RAF. More widely used was the Thornycroft Amazon 6-ton 6 × 4 chassis, with a similar Coles Mark VIII turntable-mounted. Of the latter type, 1,432 were produced, plus 388 with a longer wheel base, of which the majority went into RAF service. Crossley had a version of their IGL8 6 × 4 3-ton truck, widely used by the RAF as tenders, fitted with the lighter Coles Mark VI crane, but strangely the RAF did not acquire it in quantity and the majority built went to the Army.

A few American mobile cranes, supplied under Lend/Lease, were returned to the USA after the war. Ten years after the war was over, the RAF had an establishment on paper for a holdilng of 618 cranes, but the actual number held was 835, the majority being wartime deliveries. Versions of the AEC 0854 and Thornycroft Amazon chassis with Mark VII Coles cranes remained in service into the 70s, both then classed as 5-ton fully slewing truck-mounted cranes. Heavier cranes were in use by then, including 10-ton Coles cranes, 12½-ton Jones cranes and a 25-ton Neal.

Apart from cranes, fork-lift trucks have come much into vogue in the postwar period for both industrial and service use. The RAF uses fork-lift vehicles for stores at maintenance units and in some cases for aircraft loading.

RAF

Top left *Sheer legs in use salvaging an RE8 force-landed on a flight from Sedgeford, circa 1918* (FIN A3876G via J.T.C. Long).

Left *Solid-tyred Rapier crane of Ransomes & Rapier hoisting the Bristol Pegasus XX engine of a Vickers Wellesley, circa 1938* (MoD H385).

Top *A Coles Mark VII crane salvaging a Bristol Beaufort in the Western Desert, circa 1942-3* (W.S. Good via P.H.T. Green).

Above *Early Coles crane; this patented model was known as the Coles Derby* (Coles).

Coles Mark VII on Thornycroft Amazon short wheelbase chassis attending to a 'pranged' Mosquito NFXXX of No 54 OTU at Eastmoor early 1946 (E.F. Cheesman).

Aircraft tractors

Most of the aircraft in the early days of the RNAS and RFC could be lifted by the rear of the fuselage and manhandled to required positions, such as out of hangars on to the flying field for pilots to take over. In some cases, Crossley and Talbot tenders and other vehicles were used for towing in longer hauls, such as from a forced landing in the vicinity of an airfield where the terrain permitted a trundling recovery.

The first tractors

The introduction of the Handley Page bombers brought the need for heavy haulage tractors on airfields. In those days before concrete runways, it was realised that the normal wheeled vehicles would not be suitable for towing; particularly in wet weather when airfields might become waterlogged; and a suitable tracked vehicle was sought. It is a popular misconception that the tank, being developed at this very time, was the vehicle which introduced the caterpillar track, but this is not so. It was the firm of Hornsby at Grantham which patented a so-called chain-track tractor, but because of lack of interest in Britain, it sold the Canadian and United States rights for £4,000 to the Holt Caterpillar Company of New York, who developed the idea. Hornsby later joined Ruston Proctor, which became Ruston and Hornsby, at Lincoln, where they built caterpillar tractors to the improved Holt design. Rights were extended to another Lincoln firm, Clayton & Shuttleworth, who obtained a contract from the Ministry of Munitions to build a smaller type of tracked tractor, using the Holt principle, for aircraft towing.

The Clayton aircraft tractor came out in 1917. Following satisfactory acceptance trials, orders were given, and later increased to a total of 250. Almost simultaneously with the tractor order, Clayton's received an order for

Above *AEC chassis mounting Coles crane seen in recent use in current yellow overall finish. Station badge is displayed on the cab door* (Peter G. Cooksley).

Below *Thornycroft Amazon long wheelbase type (17AC88) on Malta in light stone finish with red primer paint prior to repainting. On the cab door the station name Ta Kali appears in a circle but this photo was taken at Luqa, circa 1949* (K.M. Robertson Collection).

Clayton aircraft tractor towing Handley Page 0/400 in France, 1918 (Real Photos 2983).

50 Handley Page 0/400 bombers. The tractors were powered by 35/40 hp Dorman Stafford engines. After the war, the majority were sold as Ministry of Munitions surplus for industrial and agricultural use. In the early 20s, a few were held at the Harlescott MT Repair Depot and the MT store at Henlow, but the only stations to retain them in use were Andover, Gosport, Grain, Martlesham Heath, Old Sarum, Spittlegate and Upavon for towing and at Halton and Manston for instructional purposes. They were officially withdrawn in 1932, but it seems that some stations retained them into the 30s.

Apart from use in France, from where 20 Clayton aircraft tractors were returned to the UK in 1919, these tractors appear to have been confined to the UK. For overseas work in the early 20s, Fordson tractors were used. This was in connection with the RAF responsibility for policing Mesopotamia, (renamed Iraq in September 1921) for which a military air mail and communication route was pioneered from Cairo to Baghdad. To survey the 850-mile route, an RAF Desert Survey Team was formed. The most difficult portion was the several hundred miles of desert to be traversed from Amman onwards, where there was no habitation and the ground consisted of patches of flinty gravel and regions of fine sandy top surface, while in other parts there were extensive ridges of black volcanic rock. It was planned that landing grounds should be selected and marked at 20-mile intervals. A convoy of eight Crossley tenders, escorted by three Rolls-Royce armoured cars, set out from Amman to meet a convoy of six Ford 'T' vans from Baghdad which was similarly surveying and marking the emergency landing strips, while DH9As of No 30 Squadron provided air contact.

The planned method of marking the emergency areas with painted petrol tins did not prove effective, but since the DH9A crews could pick out the route by tyre marks in the sand, the landing ground perimeters were simply marked by

Above *A tender towing a Brequet biplane with engine trouble back to its hangar at Farnborough in the pre-1914 days of the RFC* (MoD H1553).

Below *David Brown aircraft-towing tractor in camouflage finish with radiator grill removed, about to draw a Lancaster of No 460 Squadron at Binbrook in April 1944* (E.D. Evans via Garbett/Goulding Collection).

the vehicles circling round the area several times. In time, the tyre marks faded and renewal was necessary. A chain harrow, pulled by Crossley tenders, was tried but this proved ineffectual.

Flying Officer J.J. Lloyd-Williams, watching a Fordson tractor at work on a government farm in Iraq, came up with the idea of a ploughed furrow to mark the route. The advantage of this was that such a ploughing would turn up the darker sub-soil and so colour the route. Oliver ploughs were bought to make the furrows, with Fordson tractors to tow them. Where ploughing was impossible on the black rock areas, the route was marked in paint. The ploughed section of the route extended some 300 miles from Ziza near Amman to Ramadi near Baghdad. The Fordsons, accompanied by escorts in Crossleys and Ford Ts, took some eight weeks to complete the task, which needed renewing every six months.

Fordsons became the only tractors in the RAF during the 20s, where they were known as RAF 'Type A' tractors. In the 30s two tractor types were acquired from Straussler Mechanisation Limited of London, a Type B light and Type C heavy, called 'Sturdy' and 'Hefty' tractors respectively. When war came, the RAF placed large orders with David Brown Tractors Ltd of Meltham, Huddersfield, supplemented by others from the USA—Case, Clark (Clarktor) and Cletrac (Cleveland Tractors). Additionally, Tructors were in use. These had a Fordson 4 × 2 Thames truck chassis, modified by Allan Taylor and Reynolds to have a shortened frame with gearbox alterations and were for use in lieu of tractors.

Towing technique and markings

Large-scale aircraft dispersal and depot storage in wartime greatly increased the need for aircraft towing and the supply of tractors. Aircraft towing was carried out in two ways; by a single towing bar affixed to the aircraft tailwheel, or by a bridlegear and two attachment points. This was essential for towing on soft or uneven ground. Aircraft now normally have lug attachment points on each undercarriage leg or in the nosewheel. There is a strict drill in active airfield towing, involving an NCO in charge of the handling party, who is responsible

for surveying the area of the move, an airman in the cockpit to operate the brakes as required, one airman at each wingtip to signal any obstruction and finally the tractor driver.

In the past there have been many collisions between taxying aircraft and moving or stationary vehicles, or involving aircraft hitting vehicles on take-off and landing. Tractors have been involved more than any other class of vehicle, due to their close association with the aircraft. For this reason, tractors have carried conspicuous markings. Up to late 1937 it was usual for tractors on airfields to carry a red for danger flag. This conventional sign had the drawback that it was not consistent. The flag would droop if there was insufficient wind and its degree of visibility depended very much on the angle of viewing. An unflagging three-dimensional sign was therefore introduced, consisting of two metal discs, 2 ft in diameter, placed at right angles to each other on a pole, and generally painted bright orange, although a more contrasting colour was used in the Middle East.

By mid-1938, when operational aircraft were appearing in dark green and dark earth camouflage, the Air Ministry decreed first that tractor bonnets should have white covers to render them more conspicuous and later that they should be bright orange all over. This in no way affected the carrying of the orange discs which had to project above any part of the tractor's structure. When war came, the tractors were among the first vehicles to be given a camouflage in industrial camouflage paints.

After the Second World War, the RAF had a large surplus of tractors. To meet an establishment world-wide in 1954 of 1,768, including workshop float and reserves, the Force had a surplus of 858. Of those which were held, well over 100 were of pre-war vintage. Later in the 50s, with the impetus of the Cold War and the building of the V-bomber force, a modernisation of tractors

David and Goliath. David Brown tractor (62AN42) in white finish at Singapore, towing an ML Mark 7 trolley cooling the interior of Belfast C1 XR371 Enceladus (ANZUK 72-1773).

Above *Douglas Taskmaster heavy tractor attendant on a Victor.*

Below *Douglas Taskmaster light tractor (68AN75) towing ex-French Navy Lancaster NX611 into position at RAF Scampton's main gate* (Lincs Chronicle).

followed, and nearly 1,000 were purchased 1955–1960, to replace the ageing fleet.

While David Brown light and medium industrial tractors continued to be used, they were supplemented by Taskmaster and Tugmaster tractors from F.L. Douglas (Equipment) Limited of Cheltenham. The tracked and semi-tracked tractors used in the past were replaced, although an International B100 tracked light shovel tractor has been used in recent years. Some modern aircraft tractors in service have a separate generating set with a 75 ft cable to maintain the aircraft's electrical system while it is under tow. Added to this is a longer cable for telephonic contact between tractor driver and aircraft crew.

Specialist tractors, rollers and cutters

Fork-lift tractors have come into large scale use in the RAF since the war for depot use in maintenance units and for aircraft loading. For rough terrain, Massey Ferguson Type 65 tractors have been used and for lifts of up to ten tons, Bedford SA8 4 × 2 and Leyland 19H 6 × 4 tractors have been acquired.

A maid of all work in the Army, and also used by the RAF, has been the Eager Beaver 4,000 lb fork lift rough terrain tractor produced by the Royal Ordnance Factory, Nottingham, designed to be air portable in Hercules aircraft. Powered by a Perkins 4-236 diesel, it had a climbing ability of one in two.

From the early days of the RFC there has been a need for agricultural equipment to maintain grass airfields and airstrips, particularly rollers and grass-cutters. Local contractors were called in for this work, using their own equipment, but some equipment was taken on charge. The first mechanically driven roller acquired, recorded as 'No 1 Motor Roller, Engine Number 3652-F16, Registered Number F092' was transferred from No 66 Squadron's charge at Filton to No 56 Squadron at London Colney on August 15 1916.

In France 1914-18 the RFC/RAF was dependent on other arms of the Army for airfield maintenance, but in 1944-45 on the continent, the plant equipment of Airfield Construction Wings could soon attend to any need. Overseas between the wars, levelling and grass-cutting were normally done by native labour by hand, and at aerodromes at home the Air Ministry laid down scales of payment per acre for grass-cutting by private contractors. During the Second World War, the contract method was continued, but in East Anglia the contractors could not keep up with the demand. Equipment was bought up from sports clubs, and aircraft tractors towed the equipment on the airfields. In recent years it has been policy to let the airfield grass grow high, so discouraging birds to settle, and thereby reducing the incidence of bird strikes.

Chapter 14

Trailers and trolleys

There are two types of trailer, semi- and full. The semi-trailer rests with part of its weight on the prime mover, while the full trailer takes the full weight of its load when pulled by the prime mover. Trolleys, which in many cases are trailers, are dealt with separately.

First World War

In the early days of the RFC, the trailer was the standard vehicle for aircraft transport by road or for the retrieval of force landed aircraft. These were light two-wheel trailers, to a standard pattern, normally pulled by Crossley light tenders, and issued on a scale of four per fighter or army corps squadron.

For carrying heavy loads, such as engines, larger four-wheel 3-ton trailers came into use, and with larger aircraft, some means of transport was needed for bulky items like mainplanes. By the end of 1918, 15 different types of trailer were in use and to avoid confusion with their long descriptive names, they were retrospectively given mark numbers as follows:

Mark	*Trailer: official description*
I	Aeroplane trailer, two-wheel, pneumatic tyres, extensible platform
IA	Aeroplane trailer two-wheel pneumatic tyres, 21 ft, fixed platform
II	Hydrogen tube trailer, two-wheel with tube cradles and straps
III	Water tank trailer, two-wheel, complete with pump
IV	Silicol trailer with collapsible box body
V	Crocodile 3-ton trailer, four solid-tyred Leyland wheels (part wood)
VA	Crocodile 3-ton trailer, four solid-tyred Leyland wheels (all steel)
VI	30 cwt four-wheel trailer with pneumatic tyres, 25 ft (all steel)
VII	Steel trailer and carrier for Handley Page 0/100 or 0/400 wings, four pneumatic-tyred wheels
VIII	Handley Page wooden plane carrier, 24 ft, with capacity to carry 4 × 27 ft mainplane sections, and two ailerons or four elevators
IX	Collapsible plane carrier, 18½ ft (conversion of earlier carriers to take eight wing sections not exceeding $6\frac{2}{3}$ ft chord)
X	Collapsible plane carrier, 20½ ft (built to take ten wing sections not exceeding 7 ft chord)
XI	5-ton workshop, four-wheel, wooden construction
XIA	5-ton workshop, four-wheel, steel construction
XII	Handley Page wing trailer, 34 ft, four-wheel

Above *The Mark V crocodile 3-ton trailer, which used Leyland lorry wheels, seen here in the 1918 Lord Mayor's Show with WRAF personnel working on a DH5 fuselage* (US Signal Corps 3838).

Below *The first and basic trailer of the Royal Flying Corps, the Mark I, seen here bringing in a captured Albatros DII fuselage near Doullens in March 1917* (IWM Q1775).

Crossley tender 9017 retrieving DH9A F1631 in Iraq in the 20s (P.H.T. Green).

Between the wars

There was no shortage of trailers in the early post-war years. Apart from trailers used by depots in the UK, 669 Mark I/IA, 43 Mark II, 150 Mark III and 411 Mark IV trailers were shipped back from France between 1918-19, together with batches of specialist trailers such as 23 chart rooms. Trailers did not wear out as fast as prime movers and wartime examples remained in use until well into the 30s. A new four-wheel trailer was introduced in 1928 as the FWS Plane-carrier Type B, a new nomenclature having been introduced for vehicles and trailers using type letters. By 1939 the standard trailers were the flat two-wheel Type G, 26 ft long by 6½ ft wide (which was then about to be replaced by the similar but improved Type K), and the flat four-wheel Type I, 21½ ft long by 7½ ft wide. Largest of the trailers was the Salvage Type E, 35 ft long by 8¼ ft wide.

Up to 1927, trailers had to have rear lights at night, but from that time until 1930 reflectors were permitted under the Road Lighting Regulations. During this period the Air Ministry approved 'Reflectors bicycle red' for RAF trailers. When tail lamps were re-introduced by statute during 1930, most RAF vehicles had electric lighting systems, except on Foden and Sentinel steam engines and some large Lacre trailers which had to use oil lamps at night.

From 1931, road regulations decreed that trailers were to carry the registration of the prime mover and RAF towing vehicles from that time carried a spare metal numberplate in 16 SWG, 21 × 5½ ins with 3½ ins letters in white on a black background. Prime movers were normally restricted to one trailer, but there were some exceptions.

Second World War

From the time of the expansion preceding the war, a spate of new trailers came into use, for cooking, water supply and additional stores, as formations to support an expeditionary force were built up. To carry complete fuselages a number of articulated trailers were brought into service in 1939 with Bedford OXC or Commer Q tractive units. These large vehicles, and the 5-ton long low-loading Crossley tractive unit with Tasker trailer, introduced in October 1943, were known as 'Queen Marys' the name arising from articulated flat trailers of 35 ft having the trade name 'Queen Elizabeth'. For wing-carrying the trailers had high side girder supports. Overall, these units had a total length of 50 ft. Throughout the war the numbers of types and quantities of trailers rose steadily for various specialist tasks, not least in the field of signals.

The postwar period

Again it was a case of the relatively greater longevity of trailers to prime movers which led to wartime trailers being in use for many years of peace. Ten years after the war, the RAF held 2,962 specialist trailers, 2,105 load carrying trailers and 385 fuel and oil trailers.

A great number of trailer types are still in service. Cargo trailers used are classed by load-carrying; types in use noted within the past 20 years are as follows:

Tons	*Wheels*	*Maker/type*
½	2	Standard patterns Marks I and II
¾	2	Brockhurst (cargo)
5	2	Massey Ferguson
5	4	Eagle (dropsides)
5	4	Tasker (dropsides)
5	4	Brockhouse (flat)
6	2	Tasker (dropsides)
6	4	Eagle (flat platform)
6	8	Tasker (flat platform)
10	4	Hands (flat platform)
10	2	Tasker (semi-trailer)
10	2	Carrimore (semi-trailer)
10	2	Scammell (semi-trailer)
10	4	Tasker (semi-trailer)
15	2	Carrimore (semi-trailer)
20	8	Tasker (flat platform)

Space precludes the listing of specialist trailers, which over recent years have included domestic service vehicles such as the Brockhurst 1-ton two-wheeled sewage disposal trailer. Examples of the administrative trailers are the 10-ton Rubery Owen recruiting semi-trailer caravans, and Tasker semi-trailer class-rooms and mobile telegraph offices. The flight simulator trailer for the Gnat T1 was mounted on four-wheels by Tasker. Operational trailers account for the majority of the specialist towed vehicles. A range of four-wheel 2-ton trailers by Rubery Owen and Sentinel are used for film processing and printing. While there have never been trailer ambulances, several trailers, such as the Hands 10-ton bacteriological laboratory and decompression chamber set, have medical associations. Further examples appear in other chapters as specialist vehicles.

Trolleys

Trolleys were in use from the earliest days of the flying services for camp domestic use, and into this category could come handcarts and standard army limbers, but the concentration here is on those concerned with aircraft maintenance, whether hand or machine drawn.

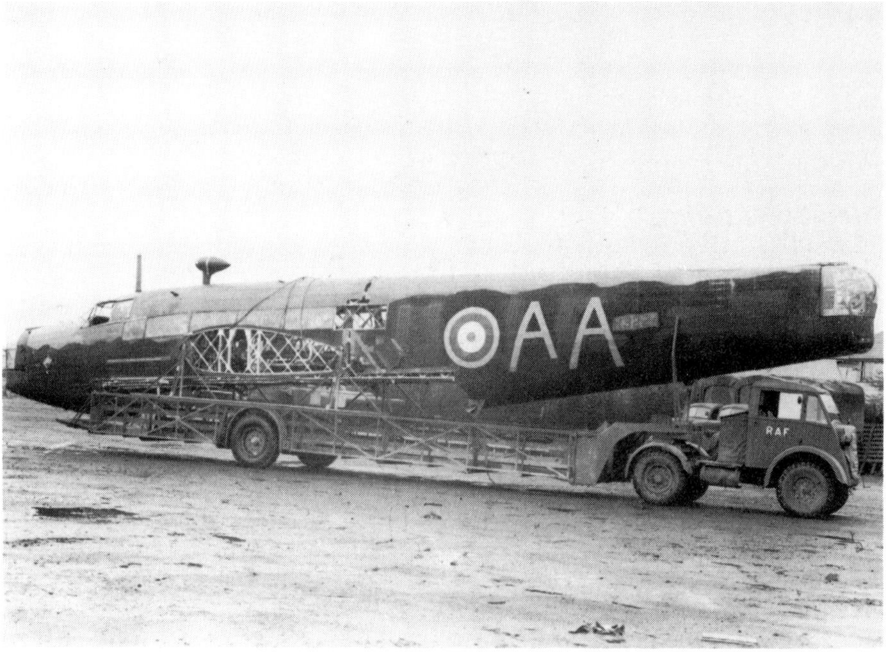

Above *Queen Mary trailer with Commer Q2 tractive unit taking Wellington 1C R3224 of No 75 Squadron for repair, after which it went to No 22 OTU and survived the war as instructional airframe 2988M* (B.J. Lowe).

Below *Queen Mary comprising Bedford tractor (42AN87) and Tasker trailer (07AU83) carrying Sunderland components for an RAF Museum exhibit, as described in the Patrick Stephens book* Epics of Aviation Archaeology. *It bears the warning at the rear 'CAUTION 52 FEET LONG'* (RAF Museum).

Old aircraft, modern semi-trailer. An AEC tractor and two-double-wheeled semi-trailer carrying a salvaged Bolingbroke from Canada to produce an example of a Blenheim type for the RAF Museum (ABN 3753).

Seaplane trolleys

The RNAS before the First World War faced the problem of maintaining flying boats and floatplanes on land, yet operating them on water, necessitating a slipway for an interchange. Since only amphibious aircraft, rare in the early years, had wheels of their own, trolleys had to be supplied with a cradle for floats or hull, to carry the whole aircraft into the water until it floated off. The process was reversed for recovery. Right up to October 1918 launching/retrieving trolleys were supplied with each seaplane. With material shortages and scarce labour, they were from that time supplied to an establishment scale of ten per floatplane flight, and five per flying boat flight.

During the First World War, seaplane trolleys were issued to standardise as much as possible as follows:

Mark	*Applicability*
IA	Standard floatplanes
IB	Non-Standard. Converted to IA
IIA	Special to type. Withdrawn 1917
IIB	F2A/F3 flying boats
IIC	Special to type. Declared obsolete 1918
IID	F5 flying boats

The vast wartime surplus allowed the issue again of one trolley per seaplane after the war, and it was not until 1934 that a replacement standard trolley was

Above *Not an amphibian, but a Short 184 floatplane using a jettisonable wheeled dolly for take-off from a carrier in late 1918* (Wing Commander L.H. Stewart OBE).

Below *Bomb train for Wellington of the Desert Air Force* (Canadian Archives RE64-15).

The Hawker Norn, of which one was built for ground handling trials, seen in 1930 using the basis of the standard wheel skid trolley, with box container raised above wheel rim level to take the skid (via Reg Longstaff).

introduced. This proved suitable for all seaplanes except Moths and Wapitis fitted with floats.

Land trolleys

While landplanes all had at least two wheels, where there was a tail-skid instead of a tail wheel there was a ground movement problem. In the First World War most fighters could be man-handled by lifting the tail and pushing the aircraft. For heavier aircraft a simple trolley was made using two aircraft wheels and a pulling handle, but to lift Handley Page bombers for turning was a hefty task for ground crews. To aid them, a two-wheel jack, the AT Mark I, was introduced in 1917 followed in 1918 by the four-wheel AT Mark II. It was 1932 before a standard tail-skid trolley was introduced for the years of peace, by which time skids were on the way out, as tailwheels became universal.

Weapon trolleys

The first bomb trolleys were improvised from aircraft wheels, but as heavier bombs came into service, low loading trolleys were needed to bring them up under bomber fuselages where the aircraft's own hoists would take over. Made by Brockhurst and Eagle, their ends were fitted with hooks and lugs so that they could be formed into a train for towing by a tractor. The various types were designated by letter.

The skeletal A type bomb trolley was in universal use before the war, then a B type with wooden shouldering was introduced temporarily for up to 1,000 lb bombs. This was replaced by the C type which became the standard wartime trolley. It was a low four-wheel chassis of 14 ft 3½ ins wheel base and 3 ft 5 ins track. Each trolley could carry the following alternative loads (in pounds) as follows: 1 × 1,900 or 2,000; 3 × 1,000; 6 × 500; 8 × 250 bombs or 2 × 1,500 mines.

To allow bombs to be rolled on to the trolley, ramps were fitted to the Type C, which then became the Mark II. To handle the introduction of 4,000 lb

WH914

Above left *Hand-hauling 1000 lb bombs on a trolley for a Canberra B2 of No 100 Squadron* (C.C.H. Cole).

Left *Nuclear weapon trolley for Blue Steel air-to-surface cruise missile under Vulcan B2 of No 617 Squadron; trolley tractor on the right.*

Above *Air ventilated suit air supply trolley Mark 3, plugged into a Vulcan. In the background, a Victor behind the David Brown tractor* (EWS 12713).

Right *Servicing tool cabinet trolley for Vickers VC10. Tool shapes were painted on the background and after servicing cabinets were checked to ensure that each painting was covered, as a check that no tools were left in the aircraft.*

Elevating loading platform for transition of loads from aircraft to lorry-loading heights. Although a civil DC-8 is involved the platform is 71AE55 of the RAF (SATA).

Cookies, a strengthened C type Mark III was produced in 1941. This differed visually because of the introduction of splashguards over the wheels. A Mark IV utility version could only carry up to 3,000 lb but the D type which followed carried up to 4,000 lb.

Large high capacity bombs were produced by bolting 4,000 lb units together to a maximum of three. To carry such sectional bombs, the Type E trolley came into use in 1942 and carried 1 × 8,000 lb, 2 × 4,000 lb or 1 × 12,000 lb bombs. An unusual feature of this 20 ft 10 ins wheelbase vehicle was its large track of 6 ft 6 ins which made the wheels protrude each side of the chassis. As this feature could cause accidents at night, luminous discs were put on the chassis sides of those remaining in use after the war.

Two standard trolleys (built later in the war, and immediately after) of sturdy construction and incorporating splashboards, were the Type F 15 ft wheelbase and 3 ft 5 ins track which took up to 8,000 lb, and the later Type G used after the war. Special Type H trolleys had to be introduced to carry the late war special bombs such as the 12,000 lb Tallboy and 22,000 Grand Slam. After the war, missiles brought further special trolleys into use.

Trolleys galore

Starter trolleys had an important part to play in the Battle of Britain, since they were manhandled from aircraft to aircraft at dispersal and later, at training units, members of the Air Training Corps acted as trolley haulers. It might be thought that heavy trolleys would come into the category of trailers, but this was not so in the case of 6- and 10-ton four-wheel salvage trolleys introduced in late 1941, which were officially classed as trolleys. Trolleys proliferated during the war and continue to do so. There are so many trolleys for servicing and for the various aircraft systems that they would fill a book on their own if they were detailed.

Chapter 15

Signals vehicles

For the uninitiated to the signals world, it should be explained that W/T, meaning wireless telegraphy, is communication by signal (such as by Morse Code signal), whereas R/T, meaning radio telephony is communication by speech. Vehicles for communication were known as 'signals vehicles' until the Second World War when the term 'radio vehicle' came into use. Signals or radio vehicles were, in the main, standard vehicles, but with cabin-type bodywork to hold equipment for various functions. Each class of vehicle was given a type number, with the prefix RVT for Radio Vehicle Type.

'Radio', it should be appreciated, was then the general service term for all types of signal equipment, whether wireless for communication or radar for detection. Similarly a radio vehicle could house wireless or radar equipment, or an associated mobile workshop or generator for the equipment, so that there was no such thing as a radar vehicle, by name. Vehicles housing radar equipment were radio vehicles. This was the result of rigid security and the name radar only came into use in the later stages of the war. It was known first as RDF (Radio Direction Finding) and later as Radiolocation, which gave nothing away further than implying wireless direction finding, which was used operationally in the First World War.

The early years

Mobile wireless transmitters and receivers first became necessary with the RAF's policing responsibilities in Iraq. As early as 1921, W/T contact was being made between DH9As and Rolls-Royce armoured cars on the move. Since the policing was in support of Iraqi ground forces, pending the establishment of an air arm of their own, RAF contact teams operated with local troops. For such cases the RAF took over and improved on a mule-mounted transmitting and receiving set known as the RAF 'cavalry pack' set.

The sets installed in Rolls-Royce armoured cars greatly hampered their use in their primary role because of the large aerial array over the turret, as well as cramping the occupants inside with the bulky equipment. There was a need for separate vehicles for use as W/T tenders, for which a Rolls-Royce chassis was adapted, carrying a T21 transmitter and Tf receiver as fitted to DH9As. Masts carried were either long Bentwood type or a six-sectioned light steel mast. There was, however, a tendency to use the W/T tender as a baggage truck, again cramping the operators.

At home, with R/T coming into general use in the mid-20s, it was mainly for the few army co-operation stations that mobile R/T stations were required and

for this purpose Crossley tenders had been fitted out in the mid-20s. Being of wartime vintage, they underwent reconditioning pending the delivery of Morris 3-ton 6 × 4 R/T tenders on order. The first batch of these new tenders was tested in the 1928 RAF training season and found wanting. They had poor cross-country performance and reception in their cabins was bad. Initially, army co-operation squadrons were allotted one Morris and one reconditioned Crossley for R/T work, pending the finding of more suitable vehicles. The bodywork of the Morris six-wheelers could be replaced for general tender work and the new Crossley 3-ton 6 × 4 with bodywork by Mann Egerton was given cross-country trials at RAE Farnborough with a 2¼-ton load. However, this vehicle too was found unsuitable and modifications delayed its introduction into service.

As a result of earlier exercises, the need had been felt for a special wireless spares-carrying vehicle to service R/T tenders in the field. A utility van of the period, a Trojan, was fitted with low pressure pneumatic tyres but it did not distinguish itself in its unaccustomed cross-country role.

The relative merits of the Crossley with RAE improvements, and the Morris six-wheeler, both using TX28 transmitters and R64 receivers, were assessed by Nos 13 and 16 (AC) Squadrons in 1931. The Crossley's R/T had a 40-mile range and over 400 miles using W/T, while the Morris had only 30 and 65 miles respectively. Both types were then committed to tropical trials and tentative production orders were placed. Both types saw service in the Second World War when signals vehicles proliferated both in types and numbers of each type, supplemented by Italian signals vehicles captured in the Western Desert. It is not possible within a single chapter to detail the various types, and this survey is

Left *Inside the cabin of a Leyland wireless maintenance lorry, 1918* (IWM Q60874).

Right *Signals trailer, Hendon 1929* (MoD H497).

Below *Rolls-Royce wireless tender with Bentwood aerial at Ramadi with No 1 Armoured Car Company vehicles, circa 1930/31* (Wing Commander S. Threapleton).

Above *A very early Early Warning aerial trailer-mounted; the Type 11 used late in the Second World War* (K.M. Robertson Collection).

Below *Type 13 aerial (RVT461A) mounted on Austin K6 3-ton 6 × 4 vehicle* (K.M. Robertson Collection).

restricted to the standard air/ground communication vehicles used in the later stages of the Second World War and field radar ground control vehicles of the same period. Some other signals vehicles for flying control are covered in Chapter 20.

Representative RVTs of the Second World War

The RVTs 100 and 150 were the standard VHF transmitting and receiving vehicles of the RAF in the Field 1943-5. Conceived and built in the mid-war years, these two vehicles had cabins mounted on 3-ton Austin 6 × 4 K6/ZB chassis, or on a Fordson WOTIA chassis. They served reliably from Normandy to the Rhine and after the war many were scrapped or stored. In 1949 a number were taken from store and towed to St Athan for complete reconditioning. The wartime T1131 transmitter was replaced by the later T1131J fully tropicalised version and the R1132 receiver of the RVT150 was replaced with a completely new R1392 receiver. In consequence these vehicles then had type numbers painted as RVT100 Mark I and RVT150 Mark II. RVT numbers continued to be allotted to new vehicles, down to 15 cwt vans, of which the RVT441 a mobile BABS van is an example.

Perhaps the signals units to use the greatest variety of vehicles were the mobile Group Control Centres (GCC), such as those which took the field in France with the Tactical Air Force in 1944. One such unit was known as AMES (Air Ministry Experimental Station—a security cover designation) Type 70. Its 30 vehicles were Austin K6 3-ton 6 × 4, Bedford OYD 3-ton 4 × 2, Crossley Q

Aerial view of a mobile aerial of No 84 Group Control Centre on the continent 1944 (E.F. Cheesman).

3-ton 4 × 4, Dodge 3-ton 4 × 2 and Fordson WOT1 3-ton 6 × 4 trucks with Radio Vehicle Type (RVT) bodywork, numbers and function as tabled.

Vehicle No	*RVT No*	*Bodywork and function*
1	480B	Cabin with two radar display units (Types 69 and 70)
2	486B	Cabin with four radar display units (Type 69)
3	475	Universal operations room with plan position indicator and height display
4	488B	Interceptions control vehicle with display units (Type 70) and two navigation tables
5	487B	As No 4 but with perspex window opposite side
6	475	Same as vehicle No 3
7	590	Office bodywork for intelligence staff
8	489A	Combat operations vehicle
9	483B	Communications vehicle
10	481A	Lower Type 14 aerial and transmitter
11	481B	Upper Type 14 aerial and transmitter
12-13	461A	Type 13 aerial and transmitter
14	436B	Radar workshop
15	419A	Cable carrier
16	436C	Radar technical office and stores
17	435F	Aerial reflector carrier
18	481C	Generator for transmitters
19-27	456	Diesel generator vehicles
28-30	—	3-ton general purpose vehicles

For operations, vehicles 4, 5, 7 and 8 faced inwards within a large (42 × 52 ft floor area) tent as the operations control, supported by the other vehicles without.

The GCC vehicles were but examples of the many radio vehicles. Depots at Chigwell and White Waltham turned out 3,200 specialist signals vehicles, 1,180 radar vehicles and 2,600 radio associated vehicles during the war. After the war, a Radio Vehicle Disposal Unit was opened at Cheddington, near Tring in Hertfordshire, but even ten years after the war the RAF had a surplus of over 1,000 radio vehicles. In fact, discounting general purpose trucks and cars, of which the RAF had 10,253 and 4,580 respectively in 1955, the 3,800 radio vehicles on strength were the most numerous in the class of specialist vehicles in the RAF.

In more recent times, 10-ton Leyland Hippos (Type 19H/7) have been used for transporting antennae and radio stations. Direction finder sets and signals offices have been mounted on Bedford 3-ton 4 × 4 trucks. Other RVTs are trailers by Brockhouse, Rubery Owen, Sankey or Tasker.

Chapter 16

Airfield clearance vehicles

To maintain the surface of airfields in the British Expeditionary Force of 1914-18 the RFC had call on the Royal Engineers and the Labour Corps, using their own and Army Service Corps transport. A standard 3-ton Leyland could take 25 cu yds of earth or 1,500 general service shovels or 1,050 RE shovels. The Independent Force, with its need for new airfields, controlled three reserve lorry parks (Nos 5, 10 and 12) by the end of 1918. These served a large body of Chinese labourers on airfield work, and German prisoners who were not allowed to work on military establishments, and were building access roads.

In the years between the wars, airfield maintenance was the responsibility of the Air Ministry Works Department operated by a civilian organisation. When war came, it was the Royal Engineers' task to provide airfields for the RAF in France and later in the Western Desert. The first winter of the war brought out a need for snow clearance on runways and snow plough attachments were hurriedly provisioned to keep selective runways clear, but in the main it was airmen diverted to shovelling who kept them open; however, there was little urgency in the 'Phoney War' period.

Airfield maintenance and repair

Urgency came with the Battle of Britain, when the cratering of airfields brought hazards in landing and take-off. Royal Engineers and the Auxiliary Military Pioneer Corps, helped by available RAF personnel and their vehicles, were used to fill in craters. At that critical time there was a shortage of lorries—particularly tippers—in the RAF, due partly to the heavy losses of vehicles in the evacuation of France. This shortage was difficult to make up from their own resources, for while the RAF had impressed aircraft the Army had been busy impressing vehicles, leaving few available other than those employed on vital war work in civil capacities. New contracts faced long delivery delays as the motor industry was being geared for the production of armoured vehicles. As a temporary measure, the Army loaned the Air Force 1,600 lorries. With a total of 2,000 cargo lorries on strength in April 1941, to carry out airfield construction and repair work, the RAF was still 1,000 vehicles below its planned establishment of this class alone. By mid-1941, only some 300 had been delivered from industry and for these there were the demands of other theatres. For example, in West Africa, where there was no civilian reserve, a large aircraft receiving post for trans-African supply was being built, as well as coastal stations for South Atlantic patrols, so some lorries had to be diverted to this area.

Fortunately MT deliveries from Lend/Lease began to arrive in quantity later

Above *Rolba rotary blower snow-clearing apparatus at Wattisham seen August 1981. A suitable British chassis not being available for the equipment, a Magirus-Deutz 4 × 4 chassis is used* (M.J.F. Bowyer).

Below *The conventional way. Refuellers with plough attachments, with early Land-Rover (RAF 37AA68) in attendance at Leuchars in the mid-50s* (K.M. Robertson Collection).

in 1941 and the MT industry was rationalised to ensure an adequate output of transport as well as armoured vehicles. In Malta, which was difficult to supply, Army, naval and civilian impressed lorries, in addition to RAF tenders, worked on aircraft maintenance as a vital priority.

The RAF had formed an Airfield Construction Service in March 1941, placing further call on tippers as well as specialist vehicles. In the field there was dual RE/RAF airfield construction, while civil contractors were used as far as possible in the UK. In the Western Desert there was a constant need to attend to airfields with equipment buried by sandstorms. Not least of the worries in North Africa was the re-preparing of airstrips after regaining territory, as it was an enemy tactic to plough airstrips up, or crater them with explosives, before they withdrew.

Sweepers and snifters

Various improved sweepers had been found necessary, to remove obstructions from airstrips, not least the removal of barbed devices dropped by enemy aircraft to rip tyres. In April 1944 the Wellingtons of a night bomber wing in Italy experienced an alarming increase in the incidence of tyre failures, so great that the supply of outer covers could not keep pace with demand. At one period 17 aircraft were unserviceable for this reason. The cause was found to be thousands of splinters of fragmentation bombs from USAAF attacks when the airfield was in enemy hands. During winter, under a covering of mud, the metal caused little inconvenience, but with fine weather the topsoil had been blown away, exposing the jagged metal pieces. Lines of personnel sweeping the airfield proved ineffective, so the wing engineering and electrical officers designed a magnetic sweeper vehicle, named on the spot a 'Snifter'. The Mark I prototype proved a failure, a second attempt, the Snifter Mark II was under-powered and another was insufficiently robust. Finally a Mark V model was put into production. This was a Jeep-towed two-wheel trailer, similar to a battery-carrying trolley, with a 24-volt, 40-amp petrol/electric set, energising coils around soft iron cores. It was necessary to stop every 100 yds for a switch-off and metal collection, but this was the basis of later vehicles, leading to the later combined vacuum/magnetic sweepers.

Snow and dust clearance

The first specialist snow blowing vehicles were acquired by the RAF in 1942, a batch of 50 to which another 80 were added during the war, but only a few of these were available to the RAF on the continent 1944-5. A different kind of clearance crisis occurred in the liberation of Europe. Nos 3207 and 3209 Airfield Construction Wings had landed in France on D + 1 (June 7 1944) and, with REs, helped to build 81 airstrips in the British operational zone in Normandy. Unforeseen was a thick brown limestone dust which coated the airfields. When whipped into the air on take-off, the dust choked the air induction systems of aero engines. Spitfires actually had to have their Merlin engines fitted with tropical filters. An urgent call went out for airfield dampers to lay the dust and some municipal horse-drawn water carts were shipped to France. They were filled by water tanker pumps and were towed over the airstrips by tractors, or other vehicles as available, to lay the dust.

In the following winter, after the advance to the Low Countries, snow and ice brought isolation to airfields which had no special vehicles capable of snow-

Above *Back to the conventional in the 80s, an RAF AEC refueller with conventional angle blade; F-16 of Royal Netherlands Air Force on the left* (Brian Goulding).

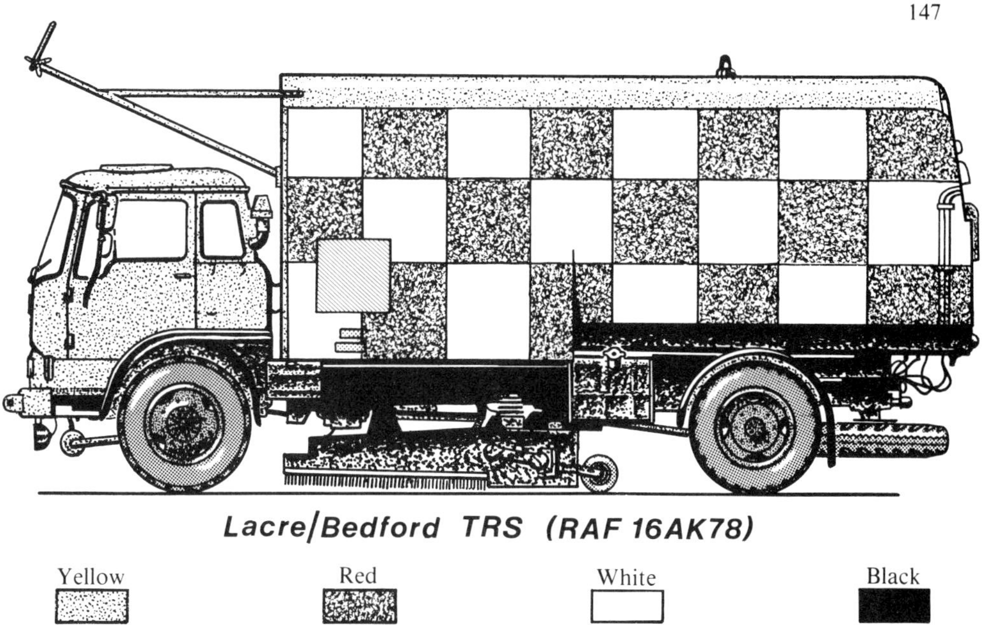

Above *Lacre/Bedford TRS (RAF 16AK78)* (Peter G. Cooksley).

Below *A TOER8 electric towtractor pulling a snow-clearing chemical spray on the runway of an RAF airfield* (Lansing Bagnall E2/7).

clearing and it was left to working parties of airmen to clear them. As a diarist at Volkel Spa in Holland put it 'No operations, no Anson so no papers, no Dakota so no mail—and no liberty truck'—for roads were as bad as the airfields.

After the war, at home, a scheme was worked out for mutual aid with local authorities who had snow ploughs. This followed the winter of 1946/47 when some communities were snow-bound for several weeks and the RAF was called upon to drop food supplies to people and cattle. With the RAF assuming responsibility for the nuclear deterrent, the possibility of a V-bomber airfield being unserviceable through snow or ice at any time was seen as a threat to national security. On March 2 1960 it was officially announced that four V-bombers could be airborne in less than four minutes from warning; this had to be under any conditions. Then there were the QRA (Quick Response Alert) airfields in Fighter (later Strike) Command and RAF Germany, equipped with Lightnings and Phantoms, alert to intercept any threat throughout 24 hours of each day. For these, a clear runway was of vital concern.

Snow- and ice-clearing has been tackled in several ways since the war. The conventional snow-plough attachment fixed to refueller vehicles was the one solution. This made it into an angle dozer which could clear an 8 ft lane of up to 12 ins of snow at five to 20 mph depending on depth of snow. No specialist vehicles were acquired in the immediate postwar years until 1953, when a dozen or so were ordered. Supplementing wartime vehicles which had been kept in service, these gave the RAF just over 150 such vehicles at the peak of the Cold War.

Since then, several types of snow-clearing vehicles have been acquired. Their purchase has not always been easy to justify in view of their very limited seasonal utilisation. A centrally held depot is not practicable, since in the conditions in which they may be required, their own mobility might be affected by road conditions. During the early 70s Douglas Bros or Steel Bros rotary Sno-Flyrs were used and other methods have since been used as illustrated. Snow-blowers with revolving spiral blades cut into frozen snow, broke it up and fed it to a heavy duty fan which blew it out into transport for removal, or piled it in a line 100 ft away.

Runway sweepers

For everyday use runway sweepers—a relatively recent acquisition for the RAF—rotary sweepers have been used, like the 1-ton Stolavis 80 or Wayne 1044 models or the large 6-ton 4 × 2 Bedford SL3. Currently, Lacre Ltd of St Albans supply their TRS (Twin-engined Runway Sweeper) to all three Services as well as to civil authorities. The name Lacre is not new to the RAF, as one of the first ever six-wheel articulated lorries, the 30 hp Lacre tractor and semi-trailer, was produced in limited numbers in the early days of the RAF, which acquired a few examples and operated them into the 20s.

The TRS sweeper by Lacre, using a Bedford chassis, can sweep efficiently up to speeds of 25 mph. A ground-hugging brush, with powered suction canopy, ingests all loose particles in a 7½ ft swathe. Solid matter is collected in a compartment with a 9,300 lb capacity. Fine grit and dust is filtered so that the outflow is practically pure air. The outrigging as shown in the sketch of a TRS is an optional fitting with wide rear-view mirrors which can be viewed through windows on the cab roof.

Chapter 17

Vehicles for gases

For the operation of airships and balloons vast quantities of gas were needed. If the bases had their own gas manufacturing plants, vehicles were needed to supply the raw materials; if not, there was the distribution of the gas under high pressure in cylinders, which needed special vehicles.

Balloons

The military balloons of the British Army at the turn of the century were operated by the Royal Engineers, whose Balloon Section then consisted of 150 officers and men, 36 horses and some 20 carts for carrying envelopes, winches, hydrogen cylinders and stores. This section later became part of the Air Battalion, Royal Engineers, from which the Royal Flying Corps was formed on April 13 1912.

By the beginning of the First World War, the Admiralty had taken over responsibility for lighter than air craft, a mistake which the British Army recognised when trench warfare set in. From May 4 1915 the French lent I Corps of the British Army an observation balloon. Four days later the first RNAS balloon section, No 2, arrived in France and was allotted to the Second Army. In June, No 4 Section arrived, followed in August by Nos 6 and 8 Sections. Shortly after this, the Army, which had lent personnel to the Navy units, took over responsibility for balloons in France, and supplied the sections with RFC tenders in place of their former naval transport.

No 4 Kite Balloon Section, RNAS—which remained in France until well into 1916—was different, in that it was completely wheel-mounted on railway trucks. To give just a couple of hours of its active existence, an extract is taken from the Section's record for March 6 1916:

'3 pm, German aeroplane circled over balloon and dropped a light (evidently a marker). Within a few minutes three shells burst close to winch siding. 3.30 pm, locomotive arrived and winch drawn out on to main line, where balloon ascended with Captain Blake RFC and Sub Lietuenant Bell as observers to altitude of 1,900 ft. 4.35 pm, Germans started shelling. Large shells burst on level near balloon. Hauled down to 800 ft. Shells followed descent and continued to burst on level with balloon. 4.50 pm, hauled down altogether. Put out no ballast, so as not to disclose bagging down position, and man-hauled winch on truck towards Poperinghe. Shelling continued very close to balloon, and ground was searched with HE (high explosive) shells. Balloon hauled down in dark with difficulty, owing to wind having changed and cable fouling the trees. Balloon also leaking and had lost her lift.'

Kite balloon section in France showing a Leyland winch lorry on the left of a handling party and Leylands behind for carrying the envelope and hydrogen cylinders (IWM Q11873).

Winches for balloons on active service had to be mobile and so were mounted, apart from the railway mounting mentioned above, on horse-drawn four-wheel carts. The winches for the balloon cable were driven by a Sykes steam engine, with vertical boiler, which turned the winding gear. Steam had to be kept up for the whole period of the balloon's ascent, ready for haul-down, which at maximum was barely 200 ft per minute. While the equipment was standing by for winch operation, the position could not be changed, as the horses could only be harnessed when the engine itself was stationary. By the end of 1918, all Sykes steam engines had been replaced by Lancaster or Panhard petrol engine winches, mounted on 3-ton vehicle chassis, giving three speeds and reverse, and achieving a 1,500 per minute haul-down.

To replace the horse-drawn vehicles, winches with motors were bought from Delahaye Motors of Paris, mounted on their 20 hp touring car chassis. These were in use from early 1916 until late 1918, but due to the firm's French Army commitments only five could be obtained. Meanwhile, in 1916, Scammel & Nephew of London, were commissioned to produce a winch of a type designed by Colonel Renard of the French Army, operated by a 20 hp Aster petrol electric engine which achieved up to a 500 ft per minute hauldown. This was mounted on a standard 3-ton Leyland chassis, but an improved model with a 35 hp engine achieving a 900 ft per minute hauldown was built in quantity for mounting on an American 4-ton 60 hp FWD or an Italian 35 hp Type Z Lancia.

With a call for kite balloons in Egypt and other minor theatres, a trial was held at Long Valley, Aldershot of the three balloon-winch-carrying chassis; the Leyland, FWD and Lancia, each loaded with two tons and taken over sand dunes and soft ground. It was the Leyland which became bogged down, the FWD by virtue of its four wheel drive proved more reliable and the Lancia faster. In consequence, from 1917, Lancia balloon winch trucks were allocated to France in place of Leylands, and FWDs were sent to the minor theatres.

Fordson Sussex balloon winch truck introduced in 1938 which served throughout the war. Immediately behind the cab is a container for the envelope when deflated and the winch mechanism has a metal grill to stop handling lines fouling the winch gear, which has a canvas cover to protect the mechanism from the elements (IWM CH1518).

During mid-1917, 24 Caquot winches driven by a 70 hp DeLion engine capable of a 1,200 ft per minute hauldown were mounted on FWD or Lancia chassis and used to replace some of the earlier winches.

Airships

Being lighter than air, airships were manhandled from their sheds, giving the impression that airships were not much involved with transport. This is a long way from the truth, for a single airship needed far more vehicle support than one aeroplane. As the RNAS Training Manual of 1915 recommends: 'A good road to the base is essential. Gas transport will be continually moving to and fro, and unless the motor lorries can approach close to the airship, an increased amount of labour will be required in moving gas tubes. The proximity of a railway station is useful, but not so essential.'

The gas used was hydrogen, the lightest form of matter known, being $\frac{1}{18}$ the weight of air, colourless, tasteless and odourless, but it was highly inflammable. Hydrogen was manufactured by the action of sodium hydroxide (caustic soda) solution on silicon. Where plants were installed at stations, special side-boarded trailers were needed, capable of being drawn by standard 3-ton tenders, to take the silicon. Hydrogen plants were established in France for the needs of the expeditionary force and at least 50 silicon trailers were on RAF strength at the time of the Armistice.

For airship sub-stations in the UK and most kite balloon locations in France, hydrogen was supplied in metal cylinders. The RNAS used Royal Navy type cylinders of 360 cu ft capacity and RAF (Royal Aircraft Factory) type produced for the Army, which had transferred its airships to the Navy on January 1 1914. All airship work was abandoned at Farnborough in February 1915 and all relevant equipment was passed on to the Navy. The RAF hydrogen cylinders of

Above *Oxygen cylinder tube trolley servicing Venom NF3 WX908 'B' of No 89 Squadron at Leconfield in the 50s* (K.M. Robertson Collection).

Left *Batteries of hydrogen cylinders on Bedford trucks, left (on 81AE08) and in the background for inflating parachute training balloons, which take 45,000 cu ft of hydrogen gas* (John Green).

Above right *Liquid oxygen container trolleys at an RAF overseas base in the 70s. Leyland Hippo refuellers in the background* (K.M. Robertson Collection).

520 cu ft capacity were 9 ft long by 9 ins diameter and weighed 2 cwt each. While empty cylinders could be returned by rail, full cylinders under railway regulations were classed as explosives and needed special transport. To get over this problem in the real age of the train, Leyland 3-ton and some other lorry types were modified to take stacked cylinders and trailers were built for the same purpose. Hydrogen cylinders were painted with red lead paint to avoid corrosion leading to leakages, plus two finishing coats in green, both as a protective and as a field camouflage.

Barrage balloon vehicles

The RAF ceased its interest in airships in 1921 and there was limited activity with kite balloons until 1940. By that time a Balloon Command had been in existence two years for the purpose of erecting a balloon barrage as an anti-aircraft measure, forcing raiding aircraft to fly high, if not actually preventing them from over-flying protected areas. The barrage brought a new range of vehicles, winch lorries and hydrogen-cylinder-carrying trailers. Three types of prime movers were used with winch-mountings, 6 × 4 3-ton British-built Dodge, Austin K6 and Fordson Sussex chassis. When war came 444 were in the London area and 180 in the provinces. Eagle trailers were fitted with metal carrying frames for the carriage of 30 cylinders each, and some Austin K6s were fitted out as re-supply oxygen cylinder carriers.

The original RAF type cylinders were retrospectively designated Type A. The two new types of cylinders used for the barrage were of medium and high pressure of 440 and 600 cu ft capacity, designated respectively types B and C. They were painted dark brown, with the valve end painted for 12 ins in signal red to identify Type B. A similar length at the end of Type C was left unpainted. In length the two types were identical and were only marginally different in diameter.

In the main the winch lorries were used at static sites, but their mobility became essential in mid-1944 when the barrage was re-deployed rapidly to meet

Just air, but it is compressed, necessitating metal cylinders and therefore a transporting trolley (Peter G. Cooksley).

the V1 flying bomb menace. Early the following year the Command was disbanded as the threat of the V1 seemed to be overcome and few German aircraft were venturing over the UK. However, it was slightly premature, as on March 3 V1s were launched against England from Holland for the first time and that night 80 German intruders, mainly Ju88s, shot down 43 RAF aircraft and destroyed 22 on the ground. However, that month did see the last of the attacks.

A few winch lorries were retained for the tethered balloon stage in the parachute training of airborne troops, a task that currently continues and for which a Leyland 19H/1 10-ton 6 × 4 truck was used in the 70s, replaced by a similar Bedford vehicle in the 80s.

Oxygen

Since 1916, the British flying services have had a need for oxygen in high altitude flying, again involving special vehicles. It can be supplied in two ways; by cylindrical tubes under pressure (like hydrogen), or as liquid oxygen in special containers. Both means need special trolleys for conveyance. The advantage of liquid oxygen was its potential, in that when the liquid was converted to a gas by heating, the original 4½-gallon jacketed vacuum containers could produce 15,400 litres of gaseous oxygen. At first, special vehicles were not available and in early 1918, in anticipation of 60 lb liquid oxygen containers taking over from cylinders, one such container was flown to France for seven days' trial. It travelled by normal road and rail transport to test its ability to withstand active service conditions. As an alternative it was proposed to supply mobile liquid oxygen producing plants. These were difficult to accommodate on a standard 3-ton tender, but by April 1918 a plant producing up to 20 litres of liquid oxygen per hour was mounted on a railway truck. However, it would appear that Stewart and Lloyd 100 cu ft storage cylinders, for which special trailers were provided, continued to be used until the end of 1918.

Between the wars, the British Oxygen Company (BOC) supplied the RAF from large capacity cylinders to fill their own 750-litre cylinders. The oxygen was supplied by BOC to ammunition parks for storage and was supplied to units by RAF transport. It was planned to have mobile oxygen plants for any expeditionary force, but these were not ready when the RAF took the field in France in 1939 and cylinders had to be sent by air in transport aircraft. During the Second World War supply was chiefly by standard cylinders. Currently both gaseous oxygen in tubes and liquid oxygen in containers have their own special trolleys for conveyance on airfields, as part of the servicing routine.

Chapter 18

RAF Regiment vehicles

The RAF Regiment, a corps which is an integral part of the RAF, was formed on February 1 1942 and by the end of that year had taken over the defence of RAF airfields in the United Kingdom. Prior to the formation of the Regiment, the RAF had operated 150 defence squadrons and 336 defence flights which were re-organised into RAF Regiment squadrons and independent flights. The Regiment was expanded for overseas tasks in the various theatres of war. For its protection duties it operated mobile anti-aircraft weapons, particularly Bofors guns, and light armoured cars which it continued to use after the war.

Lord Beaverbrook, appointed to head the newly formed Ministry of Aircraft Production in 1940, almost immediately authorised the manufacture of light armoured cars to protect factories in the event of paratroop attacks. The basis was a Standard 14B car chassis, fitted with protective armour, conceived, built and tested within a week. Named the Beaverette, after Beaverbrook, it set the style for a series of similar vehicles which were used by the Regiment in the UK. Retrospectively the first pattern became the Beaverette Mark I. This vehicle had a Standard 45 bhp four-cylinder engine which gave it a creditable road speed of 40 mph maximum in spite of its 2-ton weight. A Mark III, re-named the Beaverbug I, was fully enclosed, with a foot-operated rotating turret. Early Beaverettes were designed to take a Bren gun or a Boys anti-tank rifle. The Bugs had twin Lewis or Vickers K machine guns.

Utility vehicles were not suitable for overseas use, where the RAF's Rolls-Royce and Alvis-Straussler vehicles were past their best and understandably the Army had first call on armoured vehicles. The RAF had to take what they could get. A number of AEC Mark I armoured cars, with 2-pdr guns considered inadequate for Army use, reached No 2 AAC for trials in 1942 and the type was used by the Regiment both in the Western Desert and Egypt. Marmon Herrington armoured cars, designed by the Indianapolis firm and built by the Ford Motor Company of South Africa were acquired by the Regiment for use in the Western Desert as forward radio vehicles. More widely used by the Regiment was the Canadian-built General Motors Products Otter Mark I, a light reconnaissance armoured car with a crew of three. A number of the 1,761 produced were allotted to the RAF, but such was the Regiment's shortage of armour that they made use of captured enemy equipment, in particular German Auto-Union armoured cars in the Western Desert and Italian Lancia armoured cars in Iraq (not to be confused with the earlier Lancias discarded in the 30s).

The famous Bren-gun Carrier as it was popularly called, officially the Carrier Universal, was used by the Regiment and was also supplied to the RAF Levies in

Above *A Beaverette armoured car serving after the war at an RAF gliding school operated by Air Training Corps cadets* (K.M. Robertson Collection).

Below *Marmon Herrington armoured forward radar vehicle, with Army number, but in RAF service circa 1943* (RAF Regiment Museum).

Left *RAF Regiment personnel demonstrating man-handling a Bofors gun into position* (K.M. Robertson Collection).

Right *Introduced in the 80s, the teeth armour of the Regiment—the Scorpion armed reconnaissance vehicle* (MoD PRB 6614/32).

Below *For airfield defence the RAF Regiment was nevertheless fully mobile in its Light Anti-Aircraft role. This picture was taken of an inspection of an LAA squadron's personnel with their Bofors L70/40 guns and 3-ton trucks in the early 70s at Akrotiri, Cyprus. The Bedfords have their cab tops painted a heat-reflecting white* (K.M. Robertson Collection).

Iraq, where it continued in use until their disbandment in 1955. By early 1944 the Humber scout cars, armoured cars and light reconnaissance cars, were available in sufficient quantities to re-equip some Regiment squadrons, and form the initial equipment of those forming in the UK prior to taking the field on the invasion of the continent. The Humber scout cars saw action after the war in the Malayan emergency and it was the mid-50s before this type was withdrawn. In Aden, the RAF was issued with Ferret 1 and 2 but in general the armoured car squadrons were disbanded in the mid-50s when the Regiment took on a new image and an almost complete change of MT.

In 1959 the fire-fighting trades came under the RAF Regiment, but this range of vehicles is dealt with separately in another chapter. In 1953 the Land-Rover came into Regiment service and has become their most widely used vehicle as a carrier of personnel, machine guns and mortars. Early in 1978 there came a re-introduction of armour, with protection of Harrier sites being provided by Saladin armoured cars, Saracen armoured personnel carriers and Ferret 2 scout cars, pending the issue of modern armoured fighting vehicles. The first modern armour arrived in 1979 on loan from the Army, with RAF vehicles being delivered from August 1981.

In the anti-aircraft role, Jeeps trailed 20 mm Hispano guns. Used in the Far East, they were given up soon after the war, but the Bofors L40/70 towed gun was retained until the 70s when Bedford trucks acted as gun-tractors and ammunition carriers. These have now all been replaced by the introduction of missiles in place of projectiles—the Tigercat short range surface-to-air system in 1976 and British Aerospace's Rapier low-level air defence system from 1974, supplemented with Blindfire radar in the 80s.

Chapter 19

Wheels on rails

The RNAS, operators of armoured cars in the first months of the First World War, also operated three armoured trains, named *Churchill, Deguise* and *Jellicoe*, with the aid of the Belgian state railways. They played a notable part in the Antwerp operations of 1914. The trains were hastily assembled by the Cockerill workshops in Antwerp. Manned by RNAS officers with RNVR ratings, *Churchill*, under Commander A. Scott Littlejohns, had a hectic few days as the Germans encircled the port. Rushing around the deserted railways and sidings firing its three guns, of 6-, 4.7- and 4-ins calibre with Maxim machine-guns for close defence, it inflicted severe casualties on the advancing Germans. Because of its great mobility and self-contained ammunition supply, the train could indulge in rapid fire for short periods and move back when counter batteries came into play.

Withdrawn when the port fell, *Churchill* was brought back into action against Aubers village in the Ypres sector from early morning March 10 1915, demolishing with direct hits the church tower which was being used as an enemy observation post and sweeping the roads being used by the advancing enemy with lyddite and shrapnel. However, with trench warfare setting in, the trains were withdrawn to back areas and dismantled, though parts of one were used as a mobile balloon base as related earlier.

In the spring of 1928, during insurrection by tribes in the Southern Desert of Iraq, an armoured train prepared six years earlier under RAF control for protection of the railway, operated between Ur and Basrah. The RAF fitted the train with a T21 transmitter and Tf receiver and for internal communication between footplate and coaches a telephone system was installed. During the operation the train acted as a centre and message relaying base, but was not brought into armed action. Road/rail deployment of armour was also tried with flat trucks, hauled by a light engine of Iraqi Railways, taking a Rolls-Royce. Ramps were provided for detraining but this method does not appear to have been used operationally. Apart from deploying armour there were many instances of the RAF needing special trains for base or unit movements and some of these are recalled.

RAF rail moves

One of the first emergency rail movements of the First World War was when the Admiralty received hint of a German bombardment of the Tees and Tyne area, a month before the event. A flight of Bristol TB8s at Gosport was ordered to fly to Newcastle immediately on October 17 1914, but fog at the Farnborough stage

Top *An armoured train in effect. Rolls-Royce armoured cars and tender, truck-mounted for movement in Iraq* (Tom Hayes).

Above *An RNAS venture taken over by the RAF was the narrow gauge line at Calshot from Eaglehurst Camp to the Spit, laid down in 1917 and used for 30 years. This picture was taken in 1924* (Group Captain T.Q. Studd DFC).

Below *Saddletank 0-6-0 steam engine seen here operating on the RAF's Sleaford-Cranwell line* (MoD H2399).

Above *Masirah State Railways was the grandiose name for 3½ miles of narrow gauge track built during the Second World War on this island off the Arabian peninsula, operated by the RAF as a staging post. When the island was evacuated in 1976, one of the engines was preserved and is seen here being loaded on to an RAF vehicle.*

Below *Barclay diesel locomotive No 871* Thompson *after working for the Royal Corps of Transport, seen here pulling two FV432 armoured personnel carriers, was later transferred for work with the RAF.*

grounded the aircraft. The detachment commander, Lieutenant Christopher Draper—who in later years created a stir by flying through Tower Bridge—contacted the station master at Farnborough, who arranged flat trucks to take the three TB8 fuselages and an officer's snow-white Lancia car, while the wings were manoeuvred into long closed vans. Next morning the flight was steaming into Newcastle, and the aircraft were towed to a small airfield at Gosforth.

Aircraft for the Expeditionary Force were flown over by units, supplemented by a ferry service for reinforcements, or crated and sent by ship and rail as normal freight. There were, however, the occasional aircraft specials. The first was one arranged by the French which was shunted on to the quayside at Le Havre as the SS *Manchester Importer* came alongside on November 12, 1914. From the ship, four Avro 504s, a touring car, bombs and equipment were loaded onto the train and RNAS personnel boarded for Belfort, which was reached the following morning. There the train was hidden in a disused factory siding until nightfall. The Avros were then towed and hidden in the Belfort airship shed and made ready for the first strategic air attack in history on the Zeppelin factory at Friedrichshafen.

A large-scale RFC move came in the last two months of 1917 when two wings, their aeroplanes, kite balloons and transport were switched by rail from the Western Front to Italy. In the RAF evacuation from South Russia, a train was commandeered by the RAF in 1920. During the evacuation of France in 1940, trains were run for RAF personnel and equipment, and in the evacuation of the Dutch East Indies the RAF took over a train.

For routine movement a squadron on the move took a considerable amount of rolling stock, particularly a Handley Page bomber squadron. It took 15, later reduced to nine, trucks to take one large Hervieu III hangar. Later the railways built special trucks so that an HP squadron's hangarage, consisting of three large hangars, could, with trained packers, be fitted into three special and 12 ordinary trucks. For moves in the UK before the war there were set procedures, including bugle calls for entraining. Officers had to have 1st Class, Warrant Officers 2nd Class and NCOs and airmen 3rd Class carriages. If parties of more than 150 were to travel, the railway companies provided a special train at no additional charge than the fares.

Trucks with side supports were needed to carry wings. Fuselages were carried on bogie trucks, but if these were not available the fuselage was placed on an ordinary flat truck, shored with timber, with flat runner trucks each side to space out the overhang. Ramps were used for loading MT and even in the 20s some loading bays permitted roll-on facilities. Workshop lorries were a special case because their heavy weight, pressing down at the four wheel points, meant that the truck bottoms might need reinforcing with timbers. In all these cases the RAF was subject to railway company regulations, but the RAF also ran some railways of their own.

RAF railways

Cranwell opened in April 1916 as an RNAS station and by late 1918 it was at its peak strength as far as numbers of personnel, aircraft and vehicles were concerned. The station establishment comprised Nos 56, 57 and 58 Training Depot Stations forming the 59th Wing, with its own aeroplane repair section. There was also a Boys Training Wing and a Physical Training School, apart from the separate organisation of an airship station. During the building of this

Above *A Baguley-Drewry 0-4-0 diesel engine, one of four (Nos 3698-3701) operating on the Chilmark railway* (Derek Adamson).

Below *All-red Ruston & Hornsby diesel, Air Ministry works No 165, built in 1940, which served as a fire engine after the war until the mid-70s.* (Derek Adamson).

large camp, a single track line of over five miles was built from Sleaford station and as the camp expanded, two miles of sidings were laid and a branch line to the airship sheds.

The line continued in use in the 20s, operated by the Air Ministry, who paid the LNER £500 annually for the use of Sleaford station. But because of the decrease in RAF station population after the war, the railway showed a deficit, in spite of doubling its modest fares in 1923. It also faced competition from an omnibus company in 1925. The Air Ministry, assessing its value, and taking into account the extra MT that would be necessary if the railway closed, ceased passenger traffic from November 14 1926, but there was limited goods traffic until the late 20s.

The Royal Engineers, from which the RFC evolved, had built a military railway line into the Royal Aircraft Factory (later Establishment) at Farnborough and lines were laid from Thetford into the RFC camp there. A station was built on the Hitchin-Bedford line at Henlow Camp and was so-named. Additional sidings were added in the 20s. The line was closed under the Beeching axe. Another line was laid from Birchington in Kent in 1918 to serve RAF Manston. The line fell into disuse in the 20s, one of its last main uses being the transportation of five crated Sopwith Snipes and equipment for the re-forming of No 3 Squadron RAF at Manston on April 1 1924, although the squadron moved to Upavon within a month.

In the mid-30s, the Air Ministry operated 22 locomotives. This rose to a peak of 112 during the Second World War, about half of them on standard gauge working. Some 700 ammunition trucks were in service, mostly of 3-ton capacity and some are still in use. In 1937, to provide a bomb store, the RAF had taken over limestone quarries at Chilmark in the Vale of Wardour. Some nine miles of 23-ins gauge track were laid to the various storage sites. Two types of engines were used, diesels for general work and battery-electric locomotives for the underground storage tunnels where diesel fumes would present a health hazard. At wartime peak the five bomb storage areas in Britain were filling 104 standard gauge wagons a day, at eight tons per wagon.

On January 25 1942 an underground storage tunnel at Llanberis collapsed, blocking the main entrance and a siding, burying some of the Decauville railway track and blocking off access to 14,000 tons of bombs and depth charges. Airmen and civilians were trapped, but all escaped and bomb supply was temporarily switched to a satellite bomb dump. Fauld was not so fortunate in November 1944 when 385 × 4,000 lb bombs exploded, killing 26 civilians, RAF airmen and Italian prisoners and destroying part of the rail system.

The Chilmark depot continued after the war, served by a spur of some two miles which was joined to the main London-Exeter railway. Here three standard gauge shunting engines were normally available to haul loads to and from the depot to interchange sheds for the narrow-gauge working. A passenger coach was used for conveying workers on the site. A total of 12 locomotives was used on this railway in the mid-70s, and the railway is still functioning.

Chapter 20

Special-to-purpose and miscellaneous vehicles

Within the confines of a single volume, it is not possible to give details of all the special and miscellaneous vehicles used by the RAF in its 60-odd years. In order to cover as wide a range as possible this is a pictorial chapter with expanded captions where appropriate.

Above *A standard utility and weapon-carrying vehicle of the RAF Regiment has been the long wheelbase Land-Rover 4 × 4 ¾-ton truck with a 109 ins wheelbase as compared to the 88 ins of other versions. Introduced in 1953, it has been produced concurrently with the standard version. The RAF has used this chassis for a two-stretcher ambulance for mountain rescue teams and as an airfield crash rescue truck. Both versions have also been manufactured with left-hand drive for export and the RAF has bought both to meet the needs of stations at home and abroad. There is a larger 4 × 4 1-ton Land-Rover; a multi-purpose general service truck also used by the RAF Regiment* (K.M. Robertson Collection).

Left *The 18 mph of the RAF's first utility vehicle, the Trojan van, did not fit in with the Service's image of Bristol Bulldogs hurtling through the air at 174 mph. The utility vehicles that followed were Austin, Hillman, Morris and Standard 10-12 hp cars with tilt-covered van type bodies, which were often referred to as 'Tillies'. One of the most famous utility vehicles of all time, and certainly so during the war, was the Jeep, seen here, which proved particularly useful to the RAF in the Far East. Supply to the RAF stopped in 1945 with the cessation of Lend/Lease, when dispensation was sought to retain numbers in service. These were finally discarded in the late 40s with the arrival of the Land-Rover. This 4 × 4 vehicle was developed as FV18001 and was classed as a ¾-ton truck. The Mark 2 model appearing in 1951 had a more powerful engine and selective front-drive engagement by dog clutch. During 1971 the bodywork was redesigned as the 4 × 4 ½-ton general service Rover as shown, with special van-type body, finished in red as a Red Arrows support vehicle. The Falcons free-fall parachute display team had a similar Land-Rover (71AM19) finished in blue* (Peter G. Cooksley).

Left *Many vehicles were produced by station engineers to carry non-standard equipment to meet special needs which arose, but were not anticipated at Air Ministry level. In winter, dirt was brought into bombers on muddy flying boots. This dried and left grit in the aircraft, which could foul equipment. A powerful mobile vacuum cleaner was devised at Ludford Magna to clean aircraft, seen here at work on a No 101 Squadron Lancaster. The aircraft also has a tractor with towing bar on its tailwheel and oxygen has been brought by a tube transporter on the right* (E.D. Evans via Garbett/Goulding Collection).

Above *A flying trolley. A Bloodhound air defence missile on its transporting trolley being air-lifted in a demonstration by Bristol Belvedere HCI XG451; Fairey Gyrodyne in the background* (BAC).

Left *Like passenger airlines, the RAF has had to maintain a large number of boarding steps and slopes, which must of necessity be mobile and so are a form of vehicle. In this case the Luqa gangways were soon to be discarded for this picture shows VC10 VX108* William Rhodes-Moorhouse VC *ferrying out service families on the evacuation of Malta* (K.M. Robertson Collection).

Above *Prime mover missile carrier, the AEC Mandator 4 × 2 10-ton Blue Steel cruise missile transporter/loader* (AEC).

Below *Where there was not a control tower, a truck caravan trailer was provided, finished in red and white chequered squares; but here a Standard light van has been adapted for air control use with a large astrodome and windsock. Hudson V seen landing* (ATP808).

Above *A specialist servicing tractor-drawn vehicle, the ML Coolair Mark 7 cooling the interior of an RAF VC10 XV107* James Nicholson VC. *This is one of a range of ML ground service trolleys for air conditioning the cabins of aircraft awaiting take-off* (ML).

Right *There had been a wide range of vehicles associated with flying control. Here at Khartoum by the control tower are the vehicles typical of an outpost RAF airfield in the late 40s—left to right, a utility for the use of flying control staff, a Jeep for aircraft dispersal guidance bearing a board at the rear with the words, which can be illuminated, 'STOP' and 'FOLLOW ME' and a 15 cwt baggage truck. This view was taken from the interior of a truck towards which the crew in the foreground are making their way* (Air Ministry R1316).

Below *The vehicles and red and white chequered array of a GCA (Ground Control Approach) unit seen at West Raynham in the late 40s. Essentially, the unit was mobile, for a change of runway use meant running down the generator and radar, stowing everything external aboard the vehicles and towing the Bendix-built GCA 10-ton trailer by its Diamond T prime mover, with diesel power supply, to a new site facing the new landing direction; an operation taking some two hours. The other vehicles supply the power, direction finding unit, mobile workshop, servicing facilities, rest caravan, and standby mobile power plant in an Austin K6 truck. Later some GCA trailers were mounted on turntables to obviate the need for movement while serving a particular station* (E.K. Cole Ltd).

Above *ARAA (Aerodrome Radio Approach Aid) was once referred to as the poor man's GCA, for its beaming was narrow and only one aircraft at a time could be dealt with. It was a radio method for assisting pilots to approach an airfield in low cloud or bad visibility by talk-down by VHF to a position close to the runway for visual landing. There was a set position for the Austin K6 ARAA truck; 75 yds in from the start of the runway and 50 yds from it. As the Austin weighed 6½ tons it needed a hardstanding. Some 50 yds further up the runway is the communication aerial and power tender as seen here at Shawbury in September 1953* (K.M. Robertson Collection).

Below *Bristol Siddeley turbo-starter unit as used in 1964 for starting the Bristol Siddeley Orpheus engines of the Gnats at No 4 Flying Training School, Valley* (Bristol Siddeley).

Right *One of two Land-Rovers modified for mine-clearance/bomb disposal work. These vehicles are unique in RAF service as four-track vehicles. The sprockets that replace the normal wheels give the vehicle a maximum speed of 30 mph and the body has a ground clearance of 27½ ins. The vehicles were finished in RAF blue/grey with red wings and bonnet top. Under the number-plate appear the words in red 'RAF BOMB DISPOSAL'* (Peter G. Cooksley).

Right *An improvised power unit in use in 1944 at RAF Binbrook to give power to the gyro bomb winches for quicker turn-round when Lancasters switched to the occasional day raid* (E.D. Evans via Garbett/Goulding collection).

Below *Flight Refuelling Mark 16 hose drum unit package as supplied for the first of the V-bombers, the Vickers Valiant on its special-to-type servicing trolley. The trolley is used for functional tests and servicing with the necessary tools provided in lockers on the trolley* (Flight Refuelling Ltd).

Chapter 21

Flags, pennants and stars

It was traditional for senior officers, when mounted, to have a mounted standard bearer. When MT became the normal mode of transport, it became usual for their vehicles to display a pennant. The dark blue, red and light blue flag of Commander, Royal Flying Corps in the Field, officially approved from July 1 1917, set the style for the flying service flags of the future. The red and dark blue, the colours of the Royal Engineers, represented the RFC's origins in the Air Battalion, REs, and the light blue chosen for the new Corps was appropriate to its role in the skies.

As the RAF was born in wartime, the development of much of its ceremonial was left until after the war. It was March 24 1921 before King George V signed an Order in Council introducing the RAF Ensign, which could be flown at RAF Stations, from airships in flight and on vehicles when carrying RAF senior ranks. It was recognised that many had died or been maimed flying with the roundel motif as a national marking and it was deemed an appropriate symbol for the Ensign (seen illustrated as flown by the Chief of the Air Staff). Being based on the roundel of the period, the motif is of 5:3:1 diameter proportions, not the 3:2:1 proportions of the roundel displayed on MT since 1947. In August 1927, rank flags were introduced in the form illustrated, except that the Marshal of the RAF's flag had one thick middle red stripe and two thin stripes each side.

Distinguishing flags were normally flown from a short mast at the front of the bonnet. Since this impeded the driver's view and compromised concealment, in 1938, when the Service was becoming war conscious, miniature (9 × 4½ ins) distinguishing flags for AOCs and air ranks (ie, Air Commodore and above) were introduced. They were flown from the offside front wing, to the rear of the wing lamp. It was the intention to introduce flags for lesser ranks, but when war came, the privilege of displaying flags was limited. In May 1941 it was spelt out that serving members of the Air Council and the Inspector General would only fly their flags when visiting RAF units. AOCs were permitted flags at all times and this was extended in 1943 to all officers of Air Rank commanding stations, but only within the bounds of their main and sub-stations.

In June 1945, with the war in Europe over, new regulations were introduced and the Secretary of State for Air's flag (as illustrated) was introduced; sized 12 × 6 ins. Various changes had occurred during the war and permission to fly the RAF Ensign had been extended to certain overseas air commands. Following the formation of the RAF Regiment, a flag for the Commandant, with the Regiment's badge, was introduced. Commandants of various schools and the Provost Marshal were also authorised to fly rank flags. In 1947 the

Right *Since semaphoring with two flags was difficult from the confines of a Lancia armoured carrier (HMAC* Sirius *in this case) Morse flag signalling was used by No 1 Armoured Car Company in the early 20s* (Andrews/Liddle).

Below *Rover used by the Chief of the Air Staff with four-star plate and RAF Ensign in miniature flag. The RAF registration is in the modern style of black on a white background* (Peter G. Cooksley).

Right *Limousine for the Chief of Defence Staff who could, on rotation, be an Air Force officer. The stars here appear offset on a dark blue, red and light blue background to represent the three Services, but the car is an Army Ford with old style, white on black, registration plate* (Peter G. Cooksley).

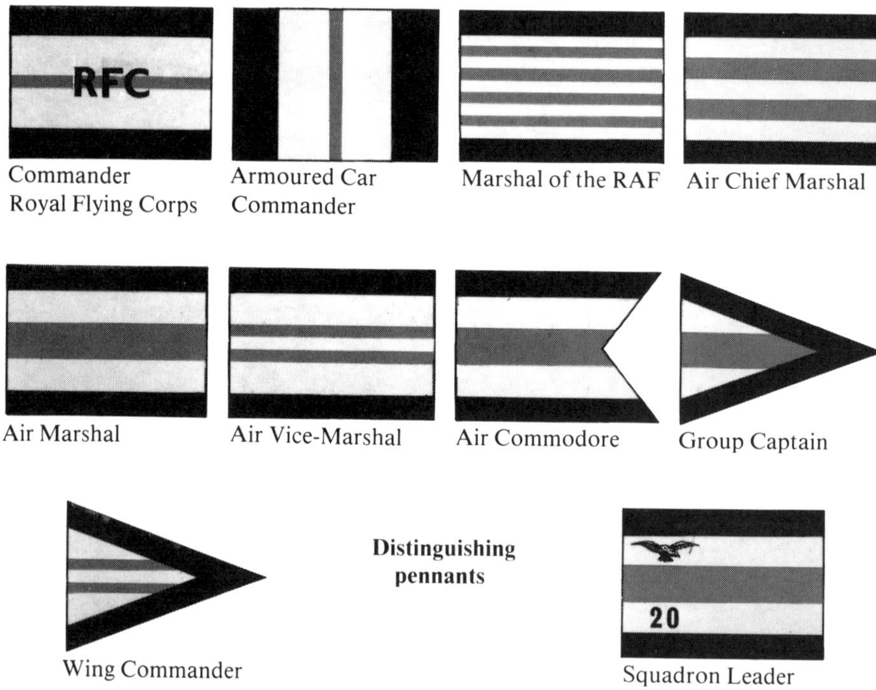

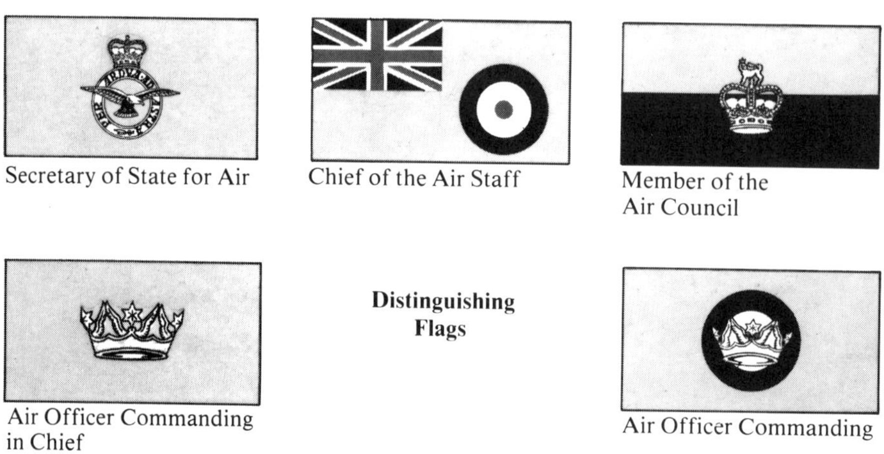

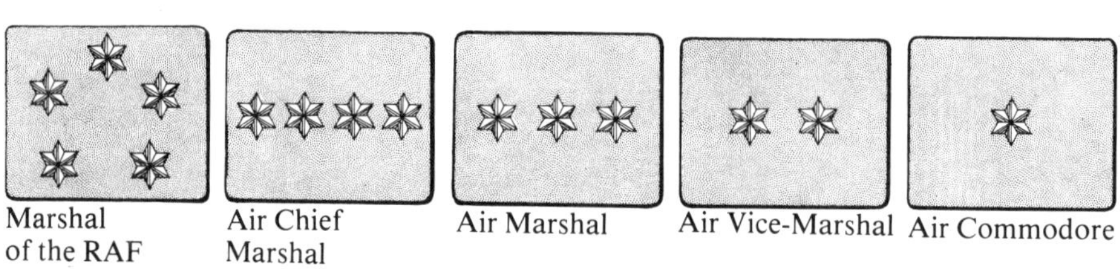

RAF stars denoting rank
(Peter G. Cooksley).

Chief of the Air Staff's flag was increased to 12 × 6 ins, while the others remained at 9 × 4½ ins, but eventually all were re-issued in the larger size, and a light blue flag with a central roundel was introduced for station commanders down to Wing Commander level. Once more the positioning reverted to one centrally above the radiator. Flags were only flown when the officer to whom they applied was actually in the vehicle. At all other times the flags were rolled around the mast and covered, or removed entirely.

The star system of notification of high ranks, as used by the USA on MT since 1917, was introduced from January 1 1951, applicable to RAF ranks as illustrated, as part of NATO standardisation. These stars supplemented RAF flags where appropriate, as they denoted rank, not the office like some of the RAF flags, and in any case the stars were positioned on front and rear bumpers. While the RAF has a light blue background for the gold stars, the Navy has dark blue and the Army a red background.

Pennants were carried on the wireless antenna of RAF Regiment armoured vehicles from the late 50s to late 60s for tactical identification in a colour code: HQ, red; 'A' Flight, yellow; 'B' Flight, green; 'C' Flight, grey and 'D' (mortar or support weapons) Flight, magenta. The Commander's vehicle of each flight carried a blue horizontal stripe on the appropriate colour. Flags in use for convoys are colour coded: blue for convoy lead vehicle, green for end of convoy, red for danger such as wide load or the carriage of explosives and an 18 × 12 ins flag with fly half diagonally in white and the remainder in a colour to denote a convoy commander.

Chapter 22

RAF MT archaeology

Examples of RAF MT, like those of its old aircraft, are fortunately preserved. A Crossley standard tender is displayed in the RAF Museum, and the Shuttleworth Collection at Old Warden, Bedfordshire, owns a 1912 six-seat Crossley Tourer presented to them by *Flight* magazine. Another Crossley tender, once used by the firm of Joseph Lucas Ltd in the 20s, is now in the hands of a private collector. Of Leyland 3-ton tenders, some six are preserved, including one currently being reconditioned. Most hardy of the survivors, and due to the vast numbers produced, some 50 Ford 'T' models are extant in Britain alone; but as the RAF usage was a mere fraction of the numbers built for civil and general military use, few are likely to be ex-RFC or RAF. Strangely, two examples of the limited build of Clayton tractors survive, one of which is loaned to the Museum of Lincolnshire Life for its agricultural association.

Most of the RAF's vehicles used between the wars, which passed the peak of their efficiency during the war, were broken up for scrap metal at a time when even personal garden railings were seized for scrap, so few remain. The surviving vehicles from the Second World War are mainly those disposed of by postwar sales arranged by the Ministry of Supply. Vehicles supplied under Lend/Lease from the USA had to be 'recaptured', to use the official American term. This either meant return to US service, or destruction to avoid causing a slump in the motor industry.

With general disarmament, there was little outlet for the RAF's armoured vehicles and as these had high scrap value, the wartime vehicles soon disappeared. However, some Beaverettes were retained and served in gliding schools, having been modified for towing. Unfortunately, these too went in the 60s just before organisations and individuals woke up to the need to preserve such relics of the past. Fortunately a private owner has a Standard Beaverette III which is now occasionally seen at displays.

A larger number of general RAF trucks, utilities and cars survive as a result of sales by the Ministry of Supply to commercial firms, and the existence of private owners who operated them for a considerable time until they came to the notice of preservation enthusiasts. However, unlike the First World War, it is difficult to point to any one type of general vehicle which is truly representative of the RAF in the Second World War, because of the supply policy of standardisation for all the Services. With the specialist vehicles it is different.

One type of fire engine practically exclusive to the RAF was the Crossley 4 × 4 fire tender of which an example is displayed in the Battle of Britain section of the RAF Museum. A Queen Mary semi-trailer is safely in the hands of

Above *One of the oldest RAF vehicles extant, the Hucks Engine Starter of the Shuttleworth collection* (Peter G. Cooksley).

Below *At the 1975 Biggin Hill Air Fair, a 1939 RAF Dennis with rear hoist, finished in the blue/grey appropriate to its day* (Peter G. Cooksley).

Above *The famous MT sheds at Farnborough, built originally as aircraft hangars, which throughout the Second World War and subsequently were used for MT, until the late 70s when they were dismantled.*

Left *Standard battery starter trolley, common to most fighter airfields 1939-45, re-emerges from the undergrowth in the late 70s* (R. Ian Wilson).

the Newark Air Museum and an Albion bowser has been preserved. A Bedford MWR signals truck, built in 1941 to act as a control truck on airfields, is kept in running order by an enthusiast at Bishop Auckland. Some owners do not appreciate the origins of their vehicles. A radar van, type not identified, with its roof antenna array removed, was noted recently on a farm at New Buckenham, Norfolk, serving as a paint store. A similar vehicle was also seen at Sandtoft, Lincolnshire. There are probably numbers of others scattered over East Anglia, Lincolnshire and Yorkshire.

Other items found on farms around the country are ex-RAF bomb trolleys, for as low trailers they provide a means of holding cattle water troughs for transport to fields as required. Of the 1,300 Commer Commando coaches built, one has been preserved, and although ex-BOAC, it is representative of a vehicle once used by RAF Transport Command.

Some vehicles after the war just disappeared; there was no official record of their scrapping and no record of where they were held. Among those untraced after the war were four Canadian-built Dodge 4 × 2 3-ton tenders, RAF Nos 176498, 176507, 185000 and 185242 discovered missing in mid-1947. In particular, whole lists of bomb trollies were reported as missing in the immediate postwar years.

In the postwar period the RAF has regularly disposed of its MT to the second-

Right *A reconstructed 12 hp Standard utility seen at King's Cross, London, 1978, with its canvas tilt removed, in a finish appropriate to 1945. This vehicle has been displayed at the Royal Scottish Museum of Flight* (R. Ian Wilson).

Overleaf *Wings to wheels.*

hand market. The Service has operated Land Rovers for 30 years and some of its early acquisitions have now become collectors items. An RAF Police ¼-ton 4 × 4 Land Rover 62AA17, discarded in 1962, was registered the following year as 9484SM and used by a builder for 12 years before it was sold as scrap to a garage where it was restored by an enthusiast from Dumfries. This is but one of many examples of ex-RAF transport on the road from continuing disposals of RAF equipment.

A different facet of RAF MT archaeology is the reserved numberplate. The prime example of this was the late Lord Brabazon of Tara, once of the RFC and RAF, who, as the holder of the first British Pilot's Licence, deserved the privilege of the registration FLY1 for his private car. This was used later by his widow. The registration 1RFC was taken up by an owner of Kingswinford, Staffordshire and RAF1 by an owner at Par, Cornwall. For those who collect car badges, there were three badges issued by RAF organisations as follows: RAF Association, RAF Air Gunners Association and RAF Prisoner of War Association.

Motor cycle enthusiasts might like to know that examples of 1917 model P&M and Triumph motor cycles are on exhibition at the RAF Museum, the latter on loan from the Science Museum. As for wartime bicycles, many may still be seen plying the roads of East Anglia.

25

Index